William Mitchell Gillespie

A treatise on levelling, topography, and higher surveying

William Mitchell Gillespie

A treatise on levelling, topography, and higher surveying

ISBN/EAN: 9783337156503

Printed in Europe, USA, Canada, Australia, Japan

Cover: Foto ©berggeist007 / pixelio.de

More available books at **www.hansebooks.com**

A TREATISE

ON

LEVELLING, TOPOGRAPHY AND HIGHER SURVEYING.

BY

W. M. GILLESPIE, LL. D.,
CIVIL ENGINEER.

EDITED BY

CADY STALEY, A. M., C. E.

NEW YORK:
D. APPLETON AND COMPANY,
549 & 551 BROADWAY.
1874.

ENTERED, according to Act of Congress, in the year 1870,
By D. APPLETON & COMPANY,
In the Office of the Librarian of Congress, at Washington.

PREFACE.

This volume, announced several years ago as being in preparation to follow the author's Land Surveying, was left at the time of his death unfinished. In preparing it for the press, the editor has endeavored to carry out, as far as possible, the plan already laid down. A considerable part of the volume has been given by the author, in the form of lectures to the civil engineering classes in Union College, and has been printed from the original manuscript.

The best authors on the subjects treated have been consulted, in order to render the work as complete as possible. The principal authorities are the following: Begat, Bourns, Breton, Chauvenet, Fenwick, Frome, Gurley, Guyot, Jackson Lee, Narrien, Puissant, Salneuve, Simms, Smith, Stephenson, Wiesbach, Williams, and the papers of the United States Coast Survey.

Wherever it has been necessary to refer to the elementary principles of surveying, reference has been made to Gillespie's Land Surveying (designated as L. S. for brevity), and the numbers of the articles referred to are enclosed in parentheses.

Union College,
Schenectady, *June*, 1870.

ANALYTICAL TABLE OF CONTENTS.

INTRODUCTION.

ART.		PAGE
1.	Levelling in General	1
2.	Direct Levelling	1
3.	Indirect Levelling	2
4.	Barometric Levelling	2
5.	Topography	2
6.	Special Objects and Difficulties	3
7.	Underground Surveying	3
8.	Water Surveying	3
9.	Reflecting Instruments	3
10.	Spherical Surveying	3
11.	Location	4

PART I.

DIRECT LEVELLING.

CHAPTER I.—GENERAL PRINCIPLES.

12.	Levelling Instruments	5
13.	Methods of Operation	5
14.	Curvature	6
15.	Refraction	7

CHAPTER II.—PERPENDICULAR LEVELS.

16.	Principle	7
17.	Plumb-line Levels	7
18.	Reflecting Levels	8

CHAPTER III.—WATER-LEVELS.

ART.		PAGE
19.	Continuous Water-levels	10
20.	Visual Water-levels	11

CHAPTER IV.—AIR BUBBLE OR SPIRIT LEVELS.

21.	The Spirit-level	12
22.	Sensibility	12
23.	Block-level	13
24.	Level with Sights	14
25.	Hand Reflected Level	14
26.	The Telescope Level	15
27.	The Y Level	16
28.	The Telescope	16
29.	The Cross-hairs	17
30.	The Level	17
31.	Supports	18
32.	Parallel Plates	18
33.	Adjustments	18
34.	First Adjustment	19
35.	Second Adjustment	20
36.	Third Adjustment	20
37.	Centring the Object-glass and Eye-piece	21
38.	Adjustment by the "Peg Method"	22
39.	Verification by another Telescope	23
40.	Egault's Level	23
41.	Troughton's Level	23
42.	Gravatt's Level	24
43.	Bourdaloue's Level	24
44.	Lenoir's Level	24
45.	Tripods	25

CHAPTER V.—RODS.

46.	How made	25
47.	Target	26
48.	Vernier's	27
49.	New-York Rod	27
50.	Boston Rod	27
51.	Speaking Rods	28

CHAPTER VI.—THE PRACTICE.

ART.		PAGE
52.	Field Routine	30
53.	Field-notes	32
54.	First Form of Field-book	33
55.	Second Form of Field-book	35
56.	Third Form of Field-book	38
57.	Best Length of Sight	39
58.	Equal Distances of Sight	39
59.	Datum-level	39
60.	Bench-marks	40
61.	Test-levels	40
62.	Limits of Precision	41
63.	Flying-levels	41
64.	Levelling for Sections	41
65.	Profiles	41
66.	Cross-levels	42

CHAPTER VII.—DIFFICULTIES.

67.	Steep Slopes	43
68.	When the Rod is too low	44
69.	When the Rod is too high	45
70.	When the Rod is too near	45
71.	Levelling across Water	45
72.	Across a Swamp or Marsh	46
73.	Through Underwood	46
74.	Over a Board Fence	46
75.	Over a Wall	46
76.	Through a House	47
77.	The Sun	47
78.	Wind	48
79.	Idiosyncrasies	48
80.	Reciprocal Levelling	48

CHAPTER VIII.—LEVELLING LOCATION.

81.	Its Nature	49
82.	Difficulties	49
83.	Staking out Work	50
84.	To Locate a Level Line	51
85.	Applications	51
86.	To run a Grade Line	52

PART II.

INDIRECT LEVELLING.

CHAPTER I.—METHODS AND INSTRUMENTS.

ART.		PAGE.
87.	Vertical Surveying	53
88.	Vertical Angles	54
89.	Instruments	55
90.	Slopes	56
91.	Theodolites	56
92.	Surveyor's Transit	56
93.	Adjustments	58
94.	Field-Work	60
95.	Angular Profiles	61
96.	Burnier's Level	62
97.	German Universal Instrument	62

CHAPTER II.—SIMPLE ANGULAR LEVELLING.

A.—For Short Distances.

98.	Principle	63
99.	Best-conditioned Angle	63

B.—For Greater Distances.

100.	Correction for Curvature	64
101.	Correcting the Result	64
102.	Correcting the Angle	64
103.	Correction for Refraction	64

C.—For Very Great Distances.

104.	Correction for Curvature	65
105.	Correction for Refraction	66
106.	Reciprocal Observation for cancelling Refraction	67
107.	Reduction to the Summits of the Signals	67
108.	When the Height of the Signal cannot be Measured	68
109.	Levelling by the Horizon of the Sea	69

CHAPTER III.—COMPOUND ANGULAR LEVELLING.

110.	By Angular Coördinates in one Plane	70
111.	By Angular Coördinates in several Planes	71
112.	Conversely	71

PART III.

BAROMETRIC LEVELLING.

CHAPTER I.—PRINCIPLES AND FORMULAS.

ART.		PAGE
113.	Principles	73
114.	Applications	73
115.	Correction for Temperature of the Mercury	74
116.	Correction for Temperature of the Air	74
117.	Other Corrections	74
118.	Rules for calculating Heights	75
119.	Formulas	75
120.	To Correct for Latitude	76
121.	Final English Formula	76
122.	French Formulas	77
123.	Babinet's Formula	77
124.	Tables	78
125.	Approximations	78

CHAPTER II.—INSTRUMENTS.

126.	Mountain Barometers	79
127.	The Aneroid Barometer	80
128.	"Boiling-point Barometer"	80
129.	Accuracy of Barometric Observations	81
130.	Simultaneous Observations	81

PART IV.

TOPOGRAPHY.

INTRODUCTION.

131.	Definition	82
132.	Systems	82

CHAPTER I.—BY HORIZONTAL CONTOUR-LINES.

133.	General Ideas	83
134.	Plane of Reference	84
135.	Vertical Distances of the Horizontal Sections	84
136.	Methods for determining Contour-lines	84

First Method.

ART.		PAGE
137.	General Method	84
138.	On a Long, Narrow Strip	85
139.	On a Broad Surface	85
140.	Surveying the Contour-lines	85
141.	Contouring with the Plane-table	86

Second Method.

142.	General Nature	86
143.	Irregular Ground	86
144.	On a Single Hill	87
145.	Extensive Topographical Survey	87
146.	Interpolation	88
147.	Interpolating with the Sector	88
148.	Ridges and Thalwegs	89
149.	Forms of Ground	90
150.	Sketching Ground by Contours	91
151.	Ambiguity	91
152.	Conventionalities	92
153.	Applications of Contour-lines	92
154.	Sections by Oblique Planes	92

CHAPTER II.—BY LINES OF GREATEST SLOPE.

155.	Their Direction	93
156.	Sketching Ground by this System	93
157.	Details of Hatchings	93

CHAPTER III.—SHADES FROM OBLIQUE AND VERTICAL LIGHT.

158.	Degree of Shade	94
159.	Shades by Tints	94
160.	Shades by Contour-lines	95
161.	Shades by Lines of Greatest Slope	95
162.	The French Method	95
163.	Lehmann's Method	95
164.	Diapason of Tints	97
165.	Shades Produced by Oblique Light	97

CHAPTER IV.—CONVENTIONAL SIGNS.

166.	Signs for Natural Surface	98
167.	Signs for Vegetation	98
168.	Signs for Water	99

ART.		PAGE
169.	Colored Topography	100
170.	Signs for Miscellaneous Objects	101
171.	Scales	103

PART V.

UNDERGROUND OR MINING SURVEYING.

172.	Objects	105

CHAPTER I.—SURVEYING AND LEVELLING OLD LINES.

173.	First Object	105
174.	The Old Method	106
175.	The New Method	108
176.	The Mining Transit	109
177.	Mapping	109

CHAPTER II.—LOCATING NEW LINES.

178.	Second Object	110
179.	When the Mine is entered by an Adit	110
180.	When the Mine is entered by a Shaft	111
181.	To Dispense with the Magnetic Needle	111
182.	Reducing Several Courses to One	112
183.	Third Object	113
184.	Problems	113

PART VI.

THE SEXTANT AND OTHER REFLECTING INSTRUMENTS.

CHAPTER I.—THE INSTRUMENTS.

185.	Principle	115
186.	Description of the Sextant	117
187.	The Box Sextant	118
188.	The Reflecting Circle	118
189.	Adjustments of the Sextant	118
190.	How to Observe	120
191.	Parallax of the Sextant	120

CHAPTER II.—THE PRACTICE.

ART.		PAGE
192.	To Set Out Perpendiculars	121
193.	The Optical Square	121
194.	To Measure a Line, one end being inaccessible	122
195.	Otherwise	124
196.	To Measure a Line when both ends are inaccessible	124
197.	Obstacles	124
198.	To Measure Heights	124
199.	Artificial Horizon	125
200.	The Sun	126
201.	Very Small Altitudes and Depressions	126
202.	To Measure Slopes	127
203.	Oblique Angles	128
204.	Advantages of the Sextant	129

PART VII.

MARITIME OR HYDROGRAPHICAL SURVEYING.

205.	Object	131

CHAPTER I.—THE SHORE LINE.

206.	The High-water Line	131
207.	The Low-water Line	132
208.	Measuring the Base	132

CHAPTER II.—SOUNDINGS.

209.	In Narrow Water	133
210.	Finding the Position of a Boat on a Sea-coast	134
211.	From the Shore	134
212.	From the Boat, with a Compass	134
213.	From the Boat, with a Sextant	134
214.	Trilinear Surveying	135
215.	Problem of the Three Points	135
216.	Instrumental Solution	137
217.	Analytical Solution	137
218.	Between Stations	138
219.	The Sounding-line	139

CHAPTER III.—TIDE-WATERS.

ART.		PAGE
220.	Tides	140
221.	Difference on Atlantic and Pacific Coast	140
222.	Mean Level of the Sea	141
223.	High and Low Water	141
224.	"Establishment" of a Place	141
225.	Tide Gauges	141
226.	Tide Tables	142
227.	Gauges in Bends	144
228.	Beacons and Buoys	144

CHAPTER VI.—THE CHART.

229.	Methods of Fixing Points on the Chart	145
230.	Conventional Signs	146

PART VIII.

SPHERICAL SURVEYING, OR GEODESY.

CHAPTER I.—THE FIELD-WORK.

231.	Nature	147
232.	Triangular Surveying	147
233.	Outline of Operations	148
234.	Measuring the Base	148
235.	Corrections of the Base	152
236.	Reducing the Base to the Level of the Sea	152
237.	A Broken Base	153
238.	Base of Verification	153
239.	Choice of Stations	154
240.	Signals	157
241.	Observations of the Angles	159
242.	Reduction to the Centre	160
243.	The Angles	162
244.	"Spherical Excess"	162
245.	Correction of the Angles	164
246.	Interior Filling-up	164

CHAPTER II.—CALCULATING THE SIDES OF THE TRIANGLES.

ART.		PAGE
247.	Methods	165
248.	Delambre's Method	165
249.	Legendre's Method	166
250.	Coördinates of the Points	166
251.	Problem I	167
252.	Second Solution	168
253.	Problem II	168
254.	Lee's Formulas	170

LEVELLING, TOPOGRAPHY, AND HIGHER SURVEYING.

INTRODUCTION.

(1.) Levelling in General. *A level surface* is one which is everywhere perpendicular to the direction of gravity, as indicated by a plumb-line, etc.; and, consequently, parallel to the surface of standing water. It is, therefore, spherical (more precisely, spheroidal), but, for a small extent, may be considered as plane. Any line lying in it is *a level line*.

A vertical line is one which coincides with the direction of gravity.

The *height* of a point is its distance from a given level surface, measured perpendicularly to that surface, and therefore in a vertical line.

LEVELLING is the art of determining the difference of the heights of two or more points.

To obtain a level surface or line, usually the latter, is the first thing required in levelling.

When this has been obtained, by any of the methods to be hereafter described, the desired height of a point may be determined *directly* or *indirectly*.

(2.) Direct Levelling. In this method of levelling, a level line is so directed and prolonged, either actually or visually, as to pass exactly over or under the point in question (i. e., so as to be in the same vertical plane with it), and the height (or depth) of the point above (or below) this level line is measured by a vertical rod, or by some similar means. The height of any other point being determined in the same manner, the difference of the two will be the height of one of the points above the other. So on, for any number of points.

DIRECT LEVELLING is the method most commonly employed. It will form Part I. of this volume.

(3.) **Indirect Levelling.** In this method of levelling the desired height is obtained by calculation from certain coördinate measured lines or angles, which fix the place of the point.

Thus, the horizontal distance from any point to a tree being known, and also the angle with the horizon made by a straight line passing from the point to the top of the tree, its height above the point can be readily calculated. This is the most simple and most usual form of this method, though many others may be employed.

INDIRECT LEVELLING will be developed in Part II.

(4.) **Barometric Levelling.** This determines the difference of the heights of two points by the difference of the weights of the portions of the atmosphere which are above each of them, as indicated by a barometer. It is explained in Part III.

(5.) **Topography.** "Surveying" determines the position of one point with reference to another, supposing them both to be situated in (or reduced to) the same level plane. "Levelling" determines how much the point in question is above or below some other level plane. Both of these combined determine where the point is "in space;" that is, where it is in reference to some known point; both *horizontally*, i. e., how far it is in front or behind, to right or to left, etc.; and *vertically*, i. e., how far it is above or below.

The position of a point in its own level plane is usually determined in "Surveying" by a pair of coördinates—lines or angles. [See L. S., (2), etc.][1] Then, its vertical distance

---

[1] L. S. will, for brevity, be used to denote the Author's "Treatise on Land Surveying," and the No. of the article referred to will be enclosed in ( ).

above or below a known level plane (i. e., its height or depth) being determined by "Levelling," becomes a "third coördinate," which fixes the place of the point.

The application of such combinations of Surveying and Levelling to determine the positions, in horizontal projection, and also the heights of the inequalities of a limited portion of the surface of the earth (its hills and hollows, ridges and valleys, etc.), is called TOPOGRAPHY. *Topographical Mapping* represents these inequalities on paper. Topography on a larger scale becomes *Geography*, properly so called. TOPOGRAPHY occupies Part IV.

(6.) Special Objects and Difficulties. The preceding methods are sufficient for the complete determination of all the features of the earth's surface; but certain operations in particular places require special methods.

(7.) Operations beneath the surface (for tunnelling, mining, etc.) being in darkness, and not easily connected with the above-ground work, involve some novel problems, and will, therefore, be treated separately in Part V., as UNDERGROUND SURVEYING AND LEVELLING.

(8.) So too, operations on the water, because of the want of steadiness in positions on its surface, require peculiar methods, and constitute another modification; described in Part VII., as WATER SURVEYING AND LEVELLING.

(9.) REFLECTING INSTRUMENTS, such as the sextant, being chiefly used in the above situation, are treated of in the preceding Part VI.

(10.) Spherical Surveying. When a great extent of country is comprised in a survey, the surface of the earth can no longer be considered as plane, but its curvature must be taken into account. Then SPHERICAL SURVEYING, or *Geodesy*, must be employed; and, instead of the straight lines and plane angles which are the coördinates of Plane Surveying, arcs of circles and spherical angles must be used. It forms Part VIII.

When still greater extents are to be surveyed, the methods of spherical surveying must be modified in accordance with the true spheroidal form of the earth.

(11.) **Location.** The name Surveying is often made to include a mode of operation which is precisely its converse.

Surveying, properly so called, determines and records the relative positions of points *as they really are*.

The converse operations have for their objects to fix the places of points *where they are desired to be*.

The term LOCATION may be extended beyond its usual meaning so as to embrace all such operations.

In laying out land, parting off portions of it, and dividing it up, the desired lines are not surveyed, but located.

In the United States Public Land Surveying, the work is almost entirely Location.

The determination of the lines of roads, their curves, etc., is *especially* Location.

The finding and pursuing a given course at sea (in Navigation) is only another form of it.

We shall find many applications of this distinction between Surveying and Location. A similar one occurs in Levelling. It should be carefully kept in mind both in "Surveying" and in "Levelling."

PART I.

DIRECT LEVELLING.

CHAPTER I.

GENERAL PRINCIPLES.

(12.) **Levelling Instruments.** The instruments employed to obtain a level line may be arranged in three classes, depending on these three principles:

1. That a line perpendicular to a vertical line is a horizontal or level line.
2. That the surface of a liquid in repose is horizontal.
3. That a bubble of air, confined in a vessel otherwise full of a liquid, will rise to the highest point of that liquid.

They will be described in the following three chapters.

(13.) **Methods of Operation.** When a level line has been obtained, by any means, the difference of heights of any two points may be found by either of these two methods:

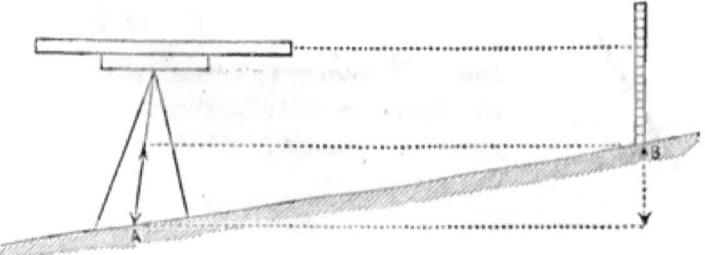

Fig. 1.

First Method.—Set the levelling instrument over one of the points, as A, in Fig. 1. Measure the height of the level line above the point. Then direct this line to a rod held on the other point, and note the reading. The difference of the two measurements at A and B will be the difference of their heights.

Second Method. Let A and B, Fig. 2, represent the two points. Set the instrument on any spot from which both the points can be seen, and at such a height that the level line will pass above the highest one. Sight to a rod held at A, and note the reading. Then turn the instrument toward B, and note the height observed on the rod held

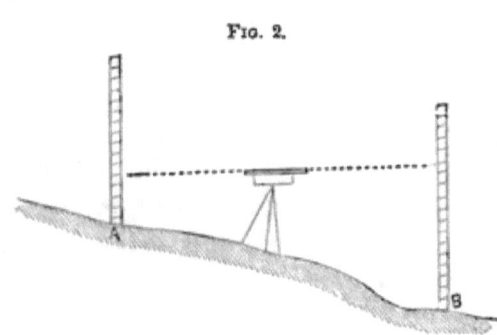

Fig. 2.

at that point. The difference of the two readings will be the difference of the heights required. The *absolute* height of the level line itself is a matter of indifference.

(14.) **Curvature.** The level line given by an instrument is tangent to the surface of the earth. Therefore, the line of *true level* is always below the line of *apparent level*. In Fig. 3, A D represents the line of apparent level, and A B the line of true level. D B is the correction for the earth's curvature. By geometry we have:

$$A D^2 = D B \times (D B + 2 B O).$$

But D B, being very small, compared with the diameter of the earth, may be dropped from the quantity in the parenthesis, and we have:

$$D B = \frac{A D^2}{2 B O}$$

i. e., the correction equals the square of the distance divided by the diameter of the earth.

The difference of height for a distance of

$$1 \text{ mile} = \frac{1}{7916} = \frac{5280 \times 12}{7916} = 8 \text{ inches.}$$

This varies as the square of the distance. The effect, if neglected, is to make distant objects appear lower than they really are.

The effect is destroyed by setting the instrument midway between the two points.

(15.) **Refraction.** Rays of light coming through the air are curved downward. The effect is, to make objects look higher than they really are. Its amount is about $\frac{1}{7}$ that of curvature, and it operates in a contrary direction.

CHAPTER II.

PERPENDICULAR LEVELS.

(16.) **Principle.** The principle upon which these are constructed is, that a line perpendicular to the direction of gravity is a level line.

(17.) **Plumb-line Levels.** The A level, Fig. 4, is so adjusted that when the plumb-line coincides with the mark on the

FIG. 4.   FIG. 5.

cross-piece, the feet of the level shall be at the same height. It is adjusted by reversion thus: Place its feet on any two points. Mark on the cross-bar the place of the plumb-line.

Turn the instrument end for end, resting it on the same points, and mark the new place of the plumb-line. The point midway between the two is the right one.

Another form is shown in Fig. 5.

The above forms are not convenient for prolonging a level line. To do this, invert the preceding form, as in Fig. 6.

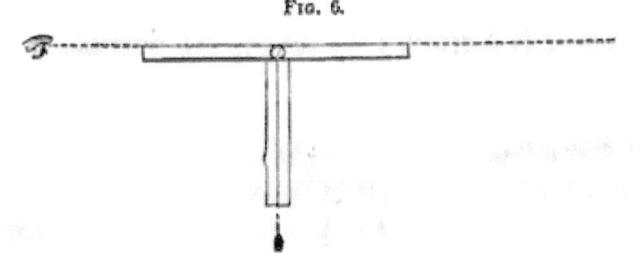

FIG. 6.

To test and adjust this, sight to some distant point nearly on a level, and mark where the plumb-line comes to on the bottom of the rod. Turn the instrument around and sight again, and note the place of the plumb-line. The midway point is the right one.

A modification of the last form is to fasten a common carpenter's square in a slit in the top of a staff, by means of a screw, and then tie a plumb-line at the angle so that it may hang beside one arm. When it has been brought to to do so, by turning the square, then the other arm will be level.

(18.) **Reflecting Levels.** In these, the perpendicular to the direction of gravity is not an actual line, but an imaginary reflected line.

It depends on the optical principle that a ray of light which meets a reflecting plane at right angles is reflected back in the same line.

When the eye sees itself in a plane mirror, the imaginary line which passes from the eye to its image is perpendicular to the mirror. Therefore, if the mirror be vertical, the line

PERPENDICULAR LEVELS.

will be horizontal. It may therefore be used like any other line of sight for determining points at the same height as itself.

The first form, Fig. 8 (Colonel Burel's), consists of a rhomb of lead, of about 2 inches on a side, and 1 inch thick.

One side (the shaded part of the figure) is faced with a mirror. The right-hand corner of the rhomb is cut off, as seen in the figure, and a wire, A B, is stretched across the mirror.

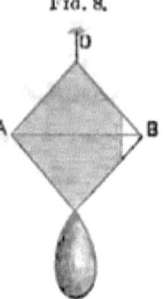

Fig. 8.

To use this, hold up the instrument, with the mirror opposite the eye, by the string D, so that the eye seems bisected in the mirror by the wire A B. Then glance through the opening at B, and any point in the line of the eye and wire will be in the same horizontal plane with them.

The correctness of the instrument may be verified in the following manner: Hold up the instrument before any plane surface, as a wall, and determine the height of some point, as previously directed. Then, without changing the height of the instrument, turn it half around, place yourself between it and the wall, and note the point of the wall which is seen in the mirror to coincide with the image of the eye.

If the two points on the wall coincide, the instrument is correct. If they do not, the mirror does not hang plumb, and the point midway between the two is the true one.

The instrument is rectified, or made to hang plumb, by means of the pear-shaped piece of lead seen attached to the lower corner of the rhomb.

 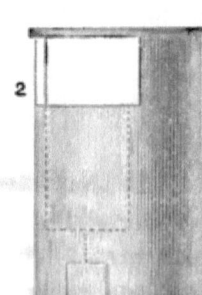

Fig. 9.

The second form consists of a hollow brass cylinder, with an opening at the upper end, as seen in Fig. 9. At the opening is a small mirror, whose vertical plane makes an angle with the vertical plane of section by which the cylinder was cut in forming the aperture. The edge of the mirror is marked thus (x) in the first half of Fig. 9. The mirror is made to hang plumb by means of a one-sided weight within the cylinder.

This is used by setting it on a stake driven into the ground, or by holding it in the hand, making the lower edge of the opening answer the same purpose as the wire in the other case.

The same methods of verification and rectification are used as with the first form of the instrument.

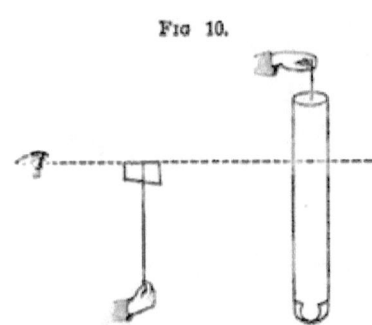

Fig 10.

The instrument, in its third form, is simply a small steel cylinder, 4″ or 5″ long, and ½″ in diameter, highly polished, and suspended from the centre of one end by a fine thread.

To use this, hold it up by the thread with one hand, and with the other hand hold a card between the eye and instrument, using the upper edge of the card, as seen reflected in the mirror, the same as the wire in the first form.

This instrument is the invention of M. Cousinery.

CHAPTER III.

WATER-LEVELS.

(19.) Continuous Water-levels. These may consist of a channel connecting the two points, and filled with water; or of a tube, usually flexible, with the ends turned up and extending from one point to the other.

By measuring up or down, from the surface of the water at each end, the relative heights of the two points may be determined.

(20.) **Visual Water-levels.** The simplest one is a short surface of water prolonged by sights at equal distances above it, as in Fig. 11.

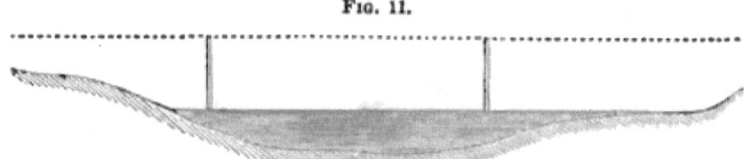

Fig. 11.

A portable form is a tube bent up at each end, and nearly filled with water. The surface of the water in one end will always be at the same height as that in the other, however the position of the tube may vary. It may be easily constructed with a tube of tin, lead, copper, etc., by bending up, at right angles, an inch or two of each end, and supporting the tube, if too flexible, on a wooden bar. In these ends, cement (with putty, twine dipped in white-lead, etc.) thin phials, with their bottoms broken off, so as to leave a free communication between them. Fill the tube and the phials, nearly to their top, with colored water. Blue vitriol or cochineal may be used for coloring it. Cork their mouths, and fit the instrument, by a steady but flexible joint, to a tripod.

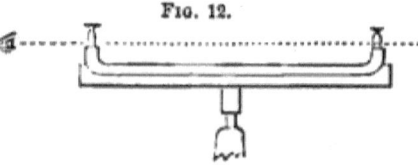

Fig. 12.

To use it, set it in the desired spot, place the tube by eye nearly level, remove the corks, and the surfaces of the water in the two phials will come to the same level. Stand about a yard behind the nearest phial, and let one eye, the other being closed, glance along the right-hand side of one phial, and the left-hand side of the other. Raise or lower the head till the two surfaces seem to coincide, and this line of sight, prolonged, will give the level line desired. Sights of equal height, floating on the water, and rising above the tops of the phials, would give a better-defined line.

CHAPTER IV.

AIR-BUBBLE OR SPIRIT LEVELS.

(21.) The "*Spirit-level*" consists essentially of a curved glass tube nearly filled with alcohol, but with a bubble of air left within, which always seeks the highest spot in the tube, and will therefore, by its movements, indicate any change in the position of the tube.

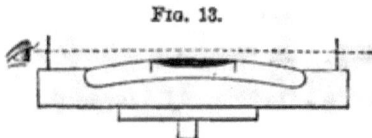

Fig. 13.

Whenever the bubble, by raising or lowering one end, has been brought to stand between two marks on the tube, or, in case of expansion or contraction, to extend an equal distance on either side of them, the bottom of the block (if the tube be in one), or sights at each end of the tube, previously properly adjusted, will be on the same level line. It may be placed on a board fixed to the top of a staff or tripod.

When, instead of the sights, a telescope is made parallel to the level, and various contrivances to increase its delicacy and accuracy are added, the instrument becomes the engineer's spirit-level.

The upper surface of the tube is usually the arc of a circle, and when we speak of lines parallel to a "level," we mean parallel to the tangent of this arc at its highest point, as indicated by the middle of the bubble.

(22.) Sensibility. This is estimated by the distance which the bubble moves for any change of inclination. It is directly proportional to the radius of curvature of the tube. To determine the radius, proceed thus:

Let S = length of the arc over which the bubble moves for an inclination of 1 second (1″).

Let R = its radius of curvature.

$$S : 2\pi R :: 1'' : 360°,$$

whence $R = 206265 \times S$,

or $S = \dfrac{R}{206265}$

S may be found by trial, the level being attached to a finely-divided vertical circle. The radius may also be found without this, thus: Bring the bubble to centre, and sight to a divided rod. Raise or lower one end of the level, and again sight to rod. Call the difference of the readings h, the distance of the rod d, and the space which the bubble moved S. Then we have two approximately similar triangles; whence $r = \dfrac{dS}{h}$.

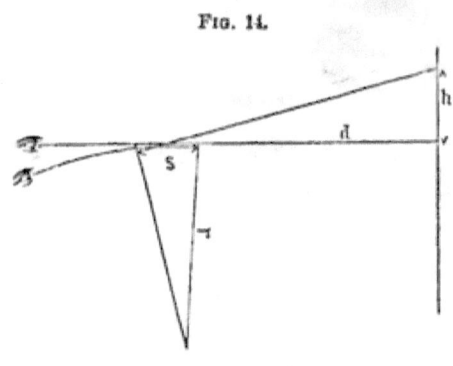

Fig. 14.

Example. At 100 feet distance, the difference of readings was 0.02 foot, and the bubble moved 0.01 foot. Then the radius was $\dfrac{100 \times 0.01}{0.02} = 50$ ft.

The sensibility of an air-bubble level equals that of a plumb-line level having a plumb-line of the same length as the radius of curvature.

(23.) **Block-level.** If this is marked by the maker, and the bubble does not come to the centre, when turned end for end, plane or grind off one end of the bottom until it does.

Fig. 15.

Otherwise, if the bubble-tube is capable of movement, raise or lower one end of it until it will verify, bringing the bubble half-way back to the middle by this means, and the other half by raising or lowering one end of the block; because the reversion has doubled the error.

Repeat this, if necessary.

Circular Level. The upper surface of this is spherical. It will therefore indicate a level in *every* direction, instead of only one, as does the preceding. It is adjusted like the last one, but in two directions, at right angles to each other.

Fig. 16.

(24.) **Level with Sights.** The line of sight is made parallel to the tangent of the level. It may be tested thus:

Fig. 17.

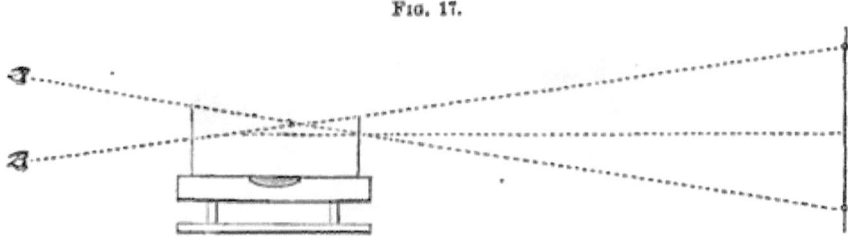

Bring the bubble to the centre of the tube and make a mark, in the line of sight, as far off as can be seen. Then turn the level end for end, and sight again. If the bubble remains in the same place, "all right." If not, rectify it by altering the sights, or by altering the marks for the bubble to come to, bringing the bubble half-way back, and trying it again.

(25.) **Hand Reflected Level.** This consists of a brass tube, about six inches long, and one inch in diameter. To the inside of the upper portion of the tube is attached a small level. A small mirror is placed at an angle in the lower side of the tube, so that it will reflect the point to which the bubble must come, in order to have the instrument level, to the eye. A small hole at one end, and a horizontal cross-hair at the other, gives the desired level line. It is used by holding it in the hand.

Fig. 18.

Fig. 18 is an approved form, made by Young, of Philadelphia. The improvement consists in the patent "Locke sight," which enables the near cross-hair to be distinctly seen at the same time as the distant object.

Fig. 19.

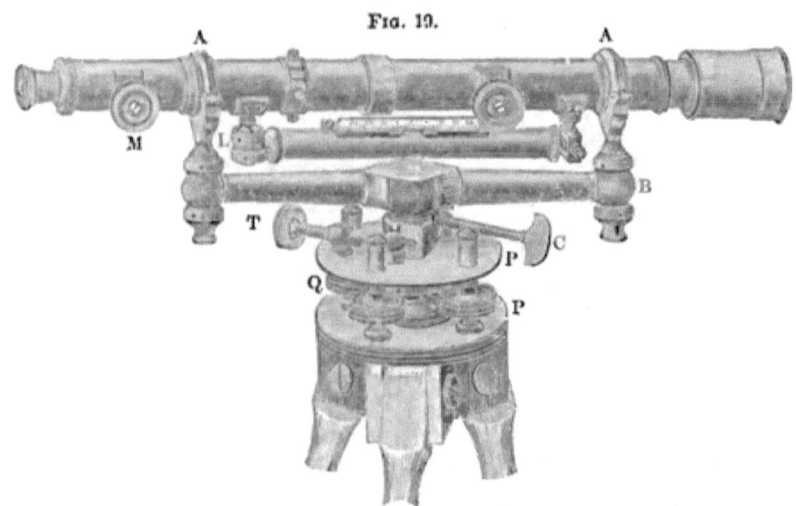

(26.) **The Telescope Level.** In this the line of collimation of the telescope corresponds to the sights of Fig. 17, and is made parallel to the level; i. e., this line is so adjusted as to be horizontal when the bubble of its level is in the centre.

There are many different forms of the Telescope Level, of which the most important ones will now be given.

NOTE.—The level, represented in Fig. 19, and described in the following articles, and the Transit, represented in Fig. 73, and described in Art. (92), are made by W. & L. E. Gurley, of Troy, N. Y., to whom the editor is indebted for valuable information respecting the construction and adjustments of the instruments.

(27.) The Y Level. This is so named from the shape of the supports of the telescope. It is the variety most used by American engineers.

Fig. 19 represents a twenty-inch Y level of the usual form. The telescope is held in the wyes by the clips, A A, which are fastened to the wyes by tapering pins, so that the telescope can be clamped in any position. The milled-headed screws at M and M are used to move the object-glass and eye-piece in and out, so as to adjust them for long and short sights, and for short-sighted and long-sighted persons. L is a spirit-level; P and P are parallel plates; C is the clamp-screw, which fastens the spindle on which the level bar, B, which supports the wyes, turns; T is the tangent screw, by which the telescope may be slowly turned around horizontally.

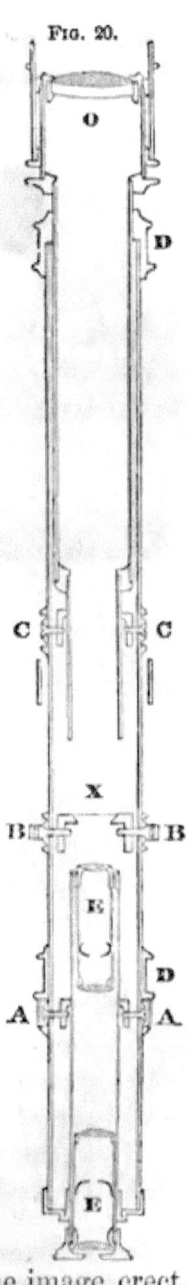

Fig. 20.

(28.) The Telescope. The arrangement of the parts of the telescope is shown in Fig. 20. O is the object-glass, by which an image of any object, toward which the telescope may be directed, is formed within the tube. E E is the eye-piece—a combination of lenses, so arranged as to magnify the small image formed by the object-glass. The cross-hairs are at X. They are moved by means of the screws shown at B B. A A are screws used for centring the eye-piece. C C are screws used for centring the object-glass. At D D are rings, or collars, of exactly the same diameter, turned very truly, by which the telescope revolves in the wyes.

The telescope shown in the figure forms the image erect. Other combinations of lenses are used, some of which invert the image; but the one here shown is generally preferred.

(29.) The Cross-hairs. These are made of very fine platinum wire or of spider-threads. They are attached to a short, thick tube, placed within the telescope-tube, through which pass loosely four screws whose threads enter and take hold of the cross-hair ring, as shown in Fig. 21.

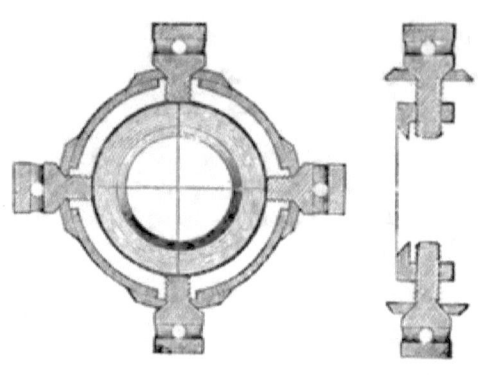

Fig. 21.

In some instruments, one of each pair of opposite screws is replaced by a spring; and the screws, instead of being capstan-headed, and moved by an "adjusting-pin," have square heads, and are moved by a "key," like a watch-key.

The line of collimation (or *line of aim*) is the imaginary line passing through the intersection of the cross-hairs and the optical centre of the object-glass.

The image formed by the object-glass should coincide precisely with the cross-hairs. When this is not the case, there will be an apparent movement of the cross-hairs, about the objects sighted to, on moving the eye of the observer. This is called *instrumental parallax.* To correct it, move the eye-piece out or in, till the cross-hairs are sharply defined against any white object. Then move the object-glass in or out, till the object is also distinctly seen. The image is now formed where the cross-hairs are, and no movement of the eye will cause any apparent motion of the cross-hairs.

(30.) The Level. This consists of a thick glass tube, slightly curved upward, and so nearly filled with alcohol that only a small bubble of air remains in the tube. This always rises to the highest part. The brass case, in which this is enclosed, is attached to the under side of the telescope, and is furnished with the means of moving, at one end vertically, and at the other, horizontally. Over the aperture, in the case, through which the bubble-phial is seen, is a graduated level-scale, numbered each way from zero at the centre.

(31.) **Supports.** The wyes in which the telescope rests, are supported by the level-bar, B, and fastened to it by two nuts at each end (one above and one below the bar), which may be moved with an adjusting-pin. The use of these nuts will be explained under "Adjustments." Attached to the centre of the level-bar is a steel spindle, made so as to turn smoothly and firmly in a hollow cylinder of bell-metal; this, again, is fitted to the main socket of the upper parallel plate.

(32.) **Parallel Plates.** It is by the aid of these that the instrument is levelled. The plates are united by a ball-and-socket joint, and are held apart by the four plate-screws, Q Q Q Q, which pass through the upper one, and press against the lower one.

To level the instrument, turn the telescope till it is brought over a pair of opposite parallel plate-screws. Then turn the pair of screws, to which the telescope has been made parallel, equally in opposite directions, screwing one in and the other out, till the bubble is brought to the centre. Then turn the telescope so as to bring it over the other pair of opposite screws, and bring the bubble to the centre, as before.

Repeat the operation, as moving one pair of screws may affect the other.

Sometimes one of each pair of opposite screws is replaced by a strong spring, and in some instruments only three screws are used.

The lower plate is screwed on to the tripod-head. For tripods, see Article (45).

(33.) **Adjustments.** The line of collimation of the telescope should be horizontal when the bubble is in the centre of the tube; which will be the case when this line is parallel to the plane of the level. But both this line and this plane are imaginary, and cannot be compared together directly. They are therefore compared indirectly. The line of collimation is made parallel to the bottom of the collars, and the plane of the level is then made parallel to them.

(34.) **First Adjustment.** *To make the line of collimation parallel to the bottoms of the collars.*

Sight to some well-defined point, as far off as it can be distinctly seen. Then revolve the telescope half around in its supports; i. e., turn it upside down. If the line of collimation was not in the imaginary axis of the rings, or collars, on which the telescope rests, it will now no longer bisect the object sighted to. Thus, if the horizontal hair was too high, as in Fig. 22, this line

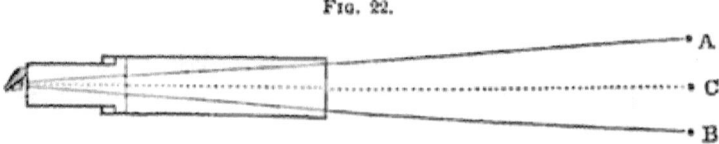

Fig. 22.

of collimation would point at first to A, and, after being turned over, it would point to B. The error is doubled by the reversion, and it should point to C, midway between A and B. Make it do so, by unscrewing the upper capstan-headed screw, and screwing in the lower one, till the horizontal hair is brought half-way back to the point. Remember that, in an erecting telescope, the cross-hairs are reversed, and *vice versa*. Bring it the rest of the way by means of the parallel plate-screws. Then revolve it in the wyes back to its original position, and see if the intersection of the cross-hairs now bisects the point, as it should. If not, again revolve, and repeat the operation till it is perfected. If the vertical hair passes to the right or to the left of the point when the telescope is turned half around, it must be adjusted in the same manner by the other pair of cross-hair screws. One of these adjustments may disturb the other, and they should be repeated alternately. When they are perfected, the intersection of the cross-hairs, when once fixed on a point, will not move from it when the telescope is revolved in its supports. This double operation is called *adjusting the line of collimation*.

It has now been brought into the centre line, or axis, of the collars, and is therefore parallel to their bottoms, or the points on which they rest, if they are of equal diameters. We have to assume this as having been effected by the maker.

In making this adjustment, the level should be clamped, but need not be levelled.

(35.) **Second Adjustment.** *To make the bottoms of the collars parallel to the plane of the level;* i. e., *to insure their being horizontal when the bubble is in the centre.*

Clamp the instrument, and bring the bubble to the centre by the parallel plate-screws. Take the telescope out of the wyes, and turn it end for end. If the bubble returns to the centre, "all right." If not, rectify it, by bringing the bubble half-way back, by means of the nuts which are above and below one end of the bubble-tube, and which work on a screw. Bring it the rest of the way by the plate-screws, and again turn end for end. Repeat the operation, if necessary.

If, in revolving the telescope (as in the first adjustment), the bubble runs toward either end, it must be adjusted sideways, by means of two screws which press horizontally against the other end of the bubble-tube. This part of the adjustment may derange the preceding part, which must, therefore, be tried again.

(36.) **Third Adjustment.** *To cause the bubble to remain in the centre of the tube when the telescope is turned around horizontally.*

To verify this, bring the bubble to the centre of the tube, and then turn the telescope half-way around horizontally. If the bubble does not remain in the centre, adjust it by bringing it half-way back by means of the nuts at the end of the level-bar. Test it by bringing it the rest of the way back by the parallel plate-screws, and again turning half-way around.

The cause of the difficulty is, that the plane of the level is not perpendicular to the axis about which it turns, and that this axis is not vertical. The above operations correct both these faults.

This adjustment is mainly for convenience, and not for accuracy, except in a very small degree.

Some instruments have no means of making the third adjustment. They must be treated thus:

Use the screws at the end of the bubble-tube, to cause the bubble to remain in the centre when the level is turned around horizontally. Then make the line of collimation parallel to the level by the method given in Art. (38), by raising or lowering the cross-hairs.

(37.) The operations of centring the eye-piece and object-glass should precede the first three which we have just explained.

Centring the Object-glass. After adjusting the line of collimation for a distant object (as explained in the "First Adjustment," Art. (34), move out the slide, which carries the object-glass, until a point ten or fifteen feet distant can be distinctly seen. Then turn the telescope half over, as before, and see if the intersection of the cross-hairs bisects the point. If not, bring it half-way back by the screws C C, Fig. 20, moving only one pair of screws at a time. Repeat the operation for a distant point, and then again for a near one, if necessary. We have now adjusted the line of collimation for long and short sights, and may assume it to be in adjustment for intermediate ones, since the bearings of the slides are supposed to be true, and their planes parallel to each other.

Centring the Eye-piece. This is to enable the observer to see the intersection of the cross-hairs precisely in the centre of the field of view of the eye-piece. It is adjusted by means of four screws, two of which are shown at A, A.

These operations are performed by the maker so permanently as to need no further attention from the engineer, and the heads of the screws, by which these adjustments are made, are covered by a thin ring which protects them from disturbance.

(**38.**) Adjustment by setting between two points, or the "*Peg Method.*" Drive two pegs several hundred feet apart, and set the instrument midway between them. Level, and sight to the rod held on each peg. The difference of the readings will be the true difference of the heights of the pegs; no matter how much the level may be out of adjustment.

Then set the level over one peg, and sight to the rod at the other. Measure the height of the cross-hairs above the first peg. The difference of this and the reading on the rod *should* equal the difference of the heights of the two points, as previously determined. If it does not, set the target to the sum or difference of the height of the cross-hairs above the first peg, and the true difference of height of the points, according as the first point is higher or lower than the second, and hold the rod on the second point. Sight to it, and raise or lower one end of the bubble-tube until the horizontal cross-hair *does* bisect the target when the bubble is in the centre. Then perform the "third adjustment."

Instead of setting *over* one peg, it is generally more convenient to set near to it, and sight to a rod held on it, and use this reading, instead of the measured height of the cross-hairs.

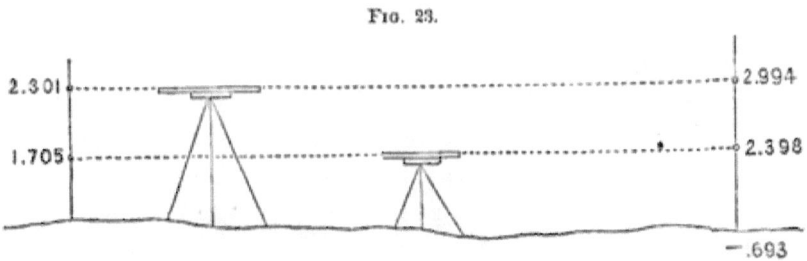

FIG. 23.

N. B. This verification should *always* be used for every level, even after the three usual adjustments have been made; for it is independent of the equality of the collars.

In running a long line of levels, let the last sight at night be taken midway between the last two "turning-point" pegs, and in the morning try their difference by setting close to the last one. This tests the level every day with very little extra labor.

(**39.**) *Verification by another Telescope.* Set up and level the instrument, and bisect the target on a distant rod. Then turn the telescope half around horizontally, and bring the bubble to the centre, if disturbed. Then take another telescope, of about the same magnifying power, but with a larger object-glass. Hold it close to the object-glass of the level, and look through the level telescope. You will see the cross-hairs plainly, and by the side of your telescope you will see the target. If the cross-hairs bisect the target, "all right." If not, adjust as in last method. If the second telescope be not larger than that of the level, hold it to one side.

(**40.**) **Egault's Level.** In this level the bubble-tube is not connected with the telescope. It is used thus:

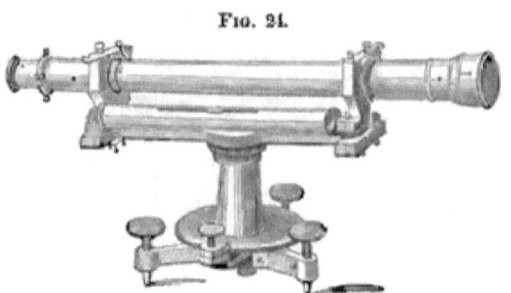

Fig. 24.

Level, and sight as usual. Then turn the telescope upside down, end for end, and half way around horizontally, and sight again. Half the sum of the two readings is the correct one, no matter how much the instrument is out of adjustment (assuming the collars to be of equal size); for the errors then cancel each other. This is the one used principally in France.

The rod used with it is marked with numbers only half the real heights above its bottom. Then the *sum* of the readings is the true one. Thus the rod itself takes the mean of the readings.

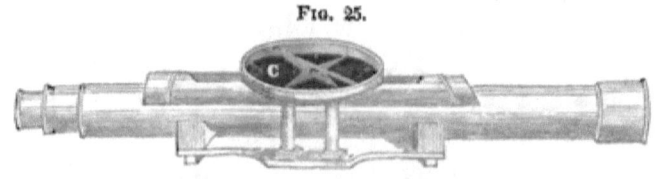

Fig. 25.

(**41.**) **Troughton's Level.** In this the bubble-tube is permanently fastened in the top of the telescope-tube. It is adjusted by the "peg method," or some similar one, the cross-hair being moved up or down until the observation gives the true difference of height of the pegs when the bubble is in the centre. Then make the "Third Adjustment," by means of the screws under the telescope.

(42.) **Gravatt's Level, or the "Dumpy Level."** Its diameter is very great, thus giving more light. Its bubble is on the top, and can be seen in a small inclined mirror, by the observer. It also has a cross-level.

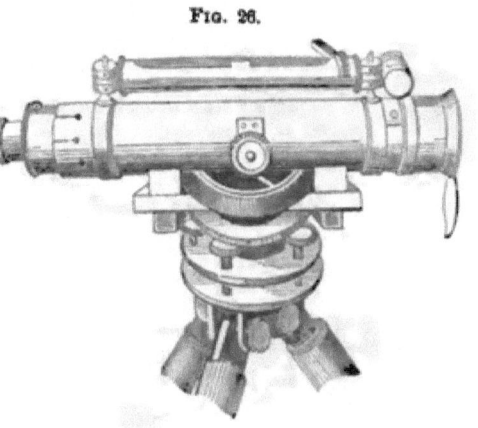

Fig. 26.

(43.) **Bourdaloue's Level.** This is a modification of Egault's. The telescope carries a steel prism near each end; one of which rests on a knife-edge, and the other on the spherical top of an adjusting-screw.

(44.) **Lenoir's Level.** In this, the telescope carries, at each end, a steel block, whose upper and lower faces are made very

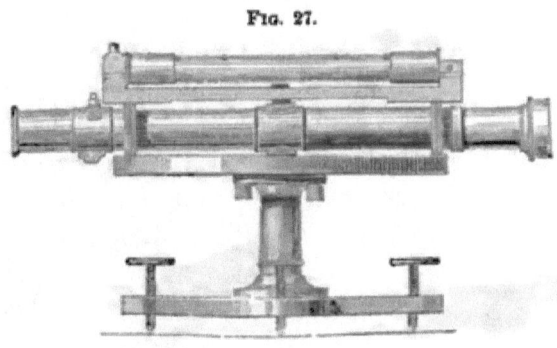

Fig. 27.

perfectly parallel. They are placed on a brass circle, which is made level by reversing a level placed upon the telescope.

(45.) **Tripods.** These consist of three legs, shod with iron, and connected by joints at the top. There are many different forms, the most common of which is given in Fig. 19. Other forms are given in Figs. 26, 28, and 29. Lightness and stiffness are the desired qualities. Of the two represented in Figs. 28 and 29, the first has the advantage of being simple and cheap; and the second of being light and yet strong.

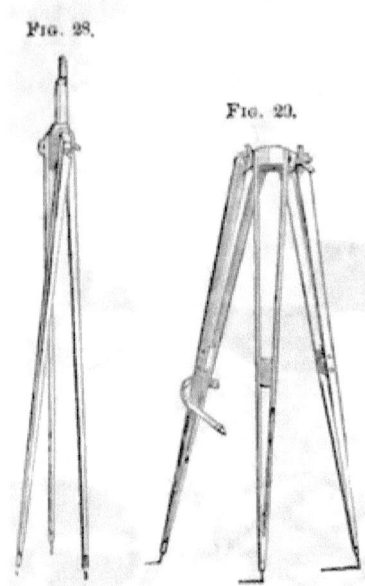

Fig. 28. Fig. 29.

Stephenson's tripod has a ball-and-socket joint below the parallel plates, so as to admit of being at once set *nearly* level on very steep slopes.

CHAPTER V.

RODS.

(46.) These should be made of light, well-seasoned wood. A plumb or level attached to them will show when they are held vertically. To detect whether the rod leans to or from the instrument, its front may be angular or curved. If angular, when held leaning toward the instrument, the lines of division will appear as in Fig. 30. When leaning from the instrument, they will appear as in Fig. 31. They are usually divided to feet, tenths, and hundredths.

(47.) Target. This is a plate of iron or brass, attached to the rod in such a way that it may be moved up and down the rod and clamped in any position. The face of the target should be painted of such a pattern that, when sighting to it, it may be very precisely bisected by the horizontal cross-hair. Some of the many varieties are given in Figs. 32–40.

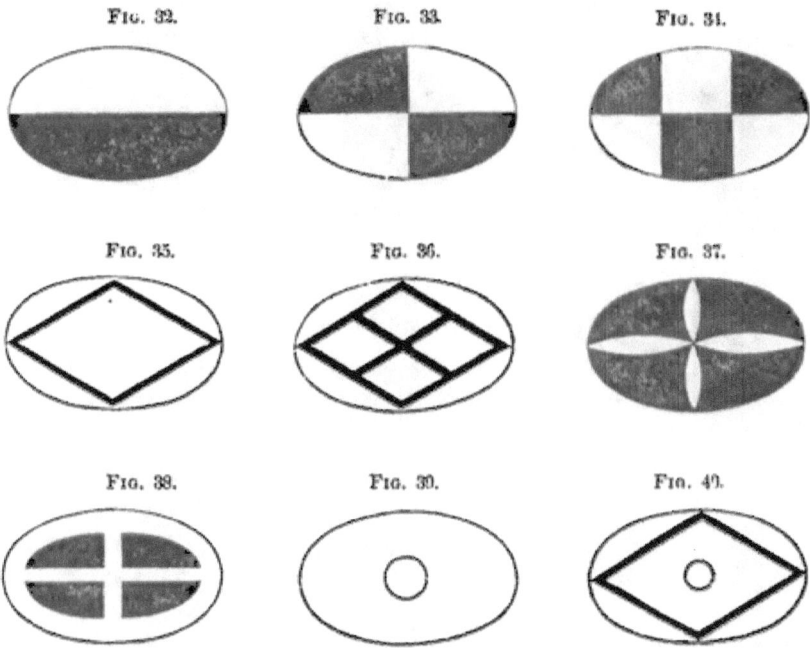

Those represented in Figs. 32, 33, and 34, are bad, because the cross-hair may be above or below the middle of the target by its full thickness, as magnified by the eye-piece of the telescope, without the error being perceptible. The next three, Figs. 35, 36, and 37, depend upon the nicety with which the eye can determine if a line bisects an angle. Fig. 38 depends upon the accuracy with which the eye can bisect a space. Fig. 39 depends upon the accuracy with which the eye can bisect a circle. Figs. 36, 37, and 40, are the best forms for use. Red and white are the best colors.

A good method of moving the target on a long rod, is by means of pulleys at the ends of the rod. A woollen cord

RODS. 27

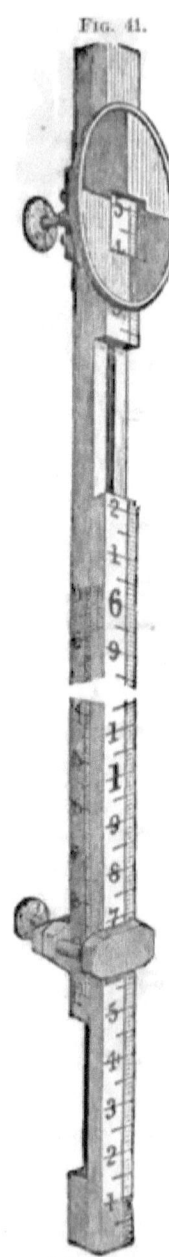

Fig. 41.

should be used, on account of its being least affected by moisture.

(48.) **Vernier's.** L. S. [343–357]. The target carries a Vernier, by which smaller spaces may be measured than those into which the rod is divided. It may be placed on the side of an aperture, in the face of the target, through which the divisions on the rod can be seen; or carried on the back or side of the rod by the target-clamp.

(49.) **The New-York Rod.** This is in two pieces, sliding one upon the other, and connected together by a tongue. It is graduated to tenths and hundredths of a foot, and can be read to thousandths by the Vernier. Up to six feet the target is used as on other rods. For greater heights, the target is fixed at $6\frac{1}{2}$ ft., and the back part of the rod, which carries the target, is shoved up (Fig. 41) until the target is bisected by the cross-hairs. Its height is then read off on the side of the rod; on which the numbers run downward, and on which is a second vernier, which gives the precise reading. It is convenient for its portability, but apt to bind or slip in sliding; i. e., to be too tight or too loose, as the weather is moist or dry.

Fig. 42.

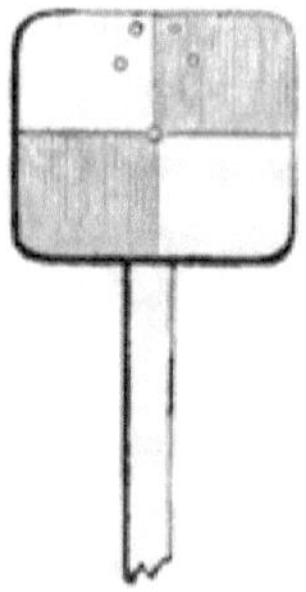

(50.) **The Boston Rod.** This is in two parts like the New York rod. The target is rectangular (Fig. 42), and is fastened to one of the pieces near its extremity. For heights less than six feet, the rod is held with the target-end down, and the target is moved up by sliding up the piece which carries it. For heights above six feet, the rod is turned end for end, bringing the target-end up, and then sliding up the piece which carries the target.

(51.) Speaking-Rods. These are rods which are read without targets, the divisions and subdivisions being painted on

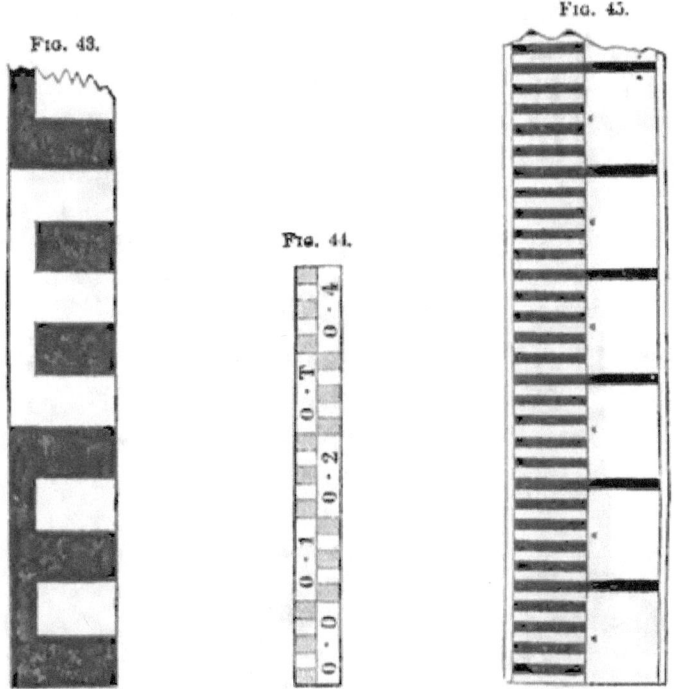

the face of the rod. They produce great saving of time and increase of accuracy.

In one form, Fig. 43, the face of the rod is divided into tenths of feet, and smaller divisions estimated.

In Bourdaloue's rod the divisions are each 4 centimetres (1.6 inches), and are numbered at half their value. He arranges them as in Fig. 44.

Gravatt's Rod, Fig. 45. This is divided to 0.01 foot. The upper hundredth of each tenth extends across the rod. Each half-tenth is marked by a dot. Each half-foot by two dots. Every other tenth is numbered, and the numbers are each

0.1 high. It is in three parts, which slide into each other like a telescope.

Barlow's Rod, Fig. 46. In this the divisions are marked by triangles, each 0.02 ft. high, so that it reads to hundredths, and less by estimation. This is based on the power the eye has in bisecting angles.

Stephenson's Rod, Fig. 47. This is based upon the principle of the Diagonal Scale. Each tenth is bisected by a horizontal line, and the diagonals enable the observer to read to hundredths.

Conybeare's Rod, Fig. 48. It reads to hundredths of a foot by means of the cross-hair bisecting the tops and bottoms and angles of hexagons. The odd tenths are made white and the even ones black. The figures are placed so that their centres are opposite the divisions they refer to.

Pemberton's Rod, Fig. 49. This is on the principle of 9

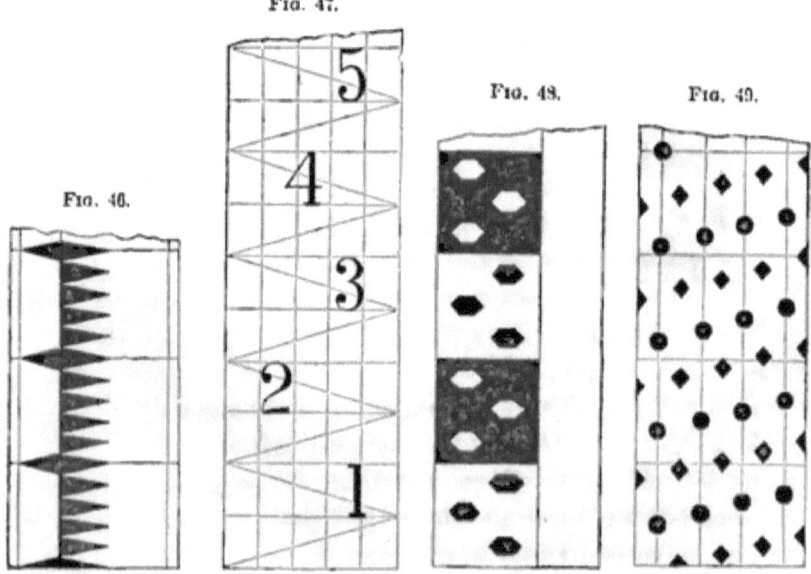

verniers placed side by side. It reads to hundredths, which are given by counting up from the dot which the hair bisects, to the dot in the same vertical line which is bisected by one of the horizontal lines which mark the tenths. The inventor claims that it can be read 9 times as far as Gravatt's.

On all speaking-rods, to avoid confounding numbers, such as 3 and 8, they may be marked thus:

1 . 2 . III . 4 . V . 6 . 7 . 8 . IX . X . 11 . XII.

The French, who go by tenths, use the following:

1 . 2 . T . 4 . V . 6 . 7 . 8. N . X.

The figures are sometimes placed with their tops on a level with the tops of the dimensions they mark, e. g., feet; and sometimes with their middles on the dividing line.

CHAPTER VI.

THE PRACTICE.

(52.) Field Routine: or, how to start and go on.

1. The rodman holds the rod on the starting-point; which may be a peg, a door-sill, or other "bench-mark" Art. (60). He stands square behind his rod, and holds it as nearly vertical as possible.

2. The leveller sets up the instrument, somewhere in the direction in which he is going, but not necessarily, or usually, in the precise line. He then levels the instrument by the parallel plate-screws, sights to the rod, and notes the reading, whether of target or speaking-rod, as a "back-sight" (B. S.), or + (plus) sight; entering it in the proper column of one of the tabular forms of field-book, given in the following articles.

3. The rodman is then sent ahead about as far as he was behind, and he there drives a "level-peg" nearly to the sur-

face of the ground, or finds a hard, well-defined point, and holds the rod upon it.

4. The leveller then again sights to the rod, and notes the reading as a "fore-sight" (F. S.), or — (minus) sight. The difference of the two readings is the difference of the heights of the points.

5. He then takes up the instrument, goes beyond the rod, any convenient distance, sets up again, and proceeds as in paragraph 2; and so on for any number of points, which will form a series of pairs. The successive observations of each pair give their difference of heights, and the combination of all these gives the difference of heights of the first and last points of the series.

6. If the vertical cross-hair be strictly vertical, it will determine whether the rod leans to the right or left. To know whether the cross-hair is vertical or not, try whether it coincides with a plumb-line; or sight to some fixed point, turn the telescope from side to side horizontally, and see if the horizontal cross-hair continues to cover the spot. If it does not, turn the telescope around in the wyes till it does; then it is truly horizontal, and the other hair, being perpendicular to it, is truly vertical. To know whether the rod leans forward or backward, have the rodman move it from and to himself. If the line bisected by the cross-hair descends in both motions, the rod was vertical. If the line rises, the rod was leaning. The lowest reading is the true one.

7. When a target is used, signals are made by the leveller with the hand, "up" and "down," to indicate in which direction to move the target. Drawing the hand to the side signifies "stop," and both hands brought together above the head signifies "all right." The rodman should move the target fast at first, and slowly after having passed the right point. When signalled "all right," he should clamp the target and show again. Then call out the reading before moving, and show it to the leveller, as either passes the other.

8. We have thus far supposed that only the difference of heights of the two extreme points is desired. But when a

section or profile of the ground is required, the rod must be held and observed, at each change of slope of the ground; or at regular distances; usually, for railroad work, at every hundred feet, and also at any change of slope between those points.

Any number of points, within sight, may have their relative heights determined at one setting of the level.

The names back-sight (B. S.) and fore-sight (F. S.) do not necessarily mean sights taken looking forward or backward (though they are generally so for turning-points), but the first sight taken, after setting up the instrument, is a B. S. or + (plus) sight, and all following ones, taken before removing the instrument, are F. S.'s, or − (minus) sights. The full meaning of this will appear in considering the forms of field-book.

All but the first and last points sighted to are called *intermediate points*, or "*intermediates*." The last point sighted to before moving the instrument is called a *turning-point* or *changing-point*.

The first and last sights, taken at any one setting of the instrument, require the greatest possible accuracy. The intermediate points may be taken only to the nearest tenth, or hundredth at most; because any error in them will not affect the final result, but only the height of that single point at which it was taken.

Two rodmen are often used to save the time of the leveller. Then it is well to use a target-rod for the "turning-points," which are often distant and need most precision, and a speaking-rod for the intermediate points. Where one rod is used, the rodman should keep notes of the readings at the turning-points.

(53.) **Field-notes.** The beginner may sketch the heights and distances measured, in a profile or side view, as in Fig. 50. But when the observations are numerous, they should be placed in one of the tabular forms given on the following pages.

FIG. 50.

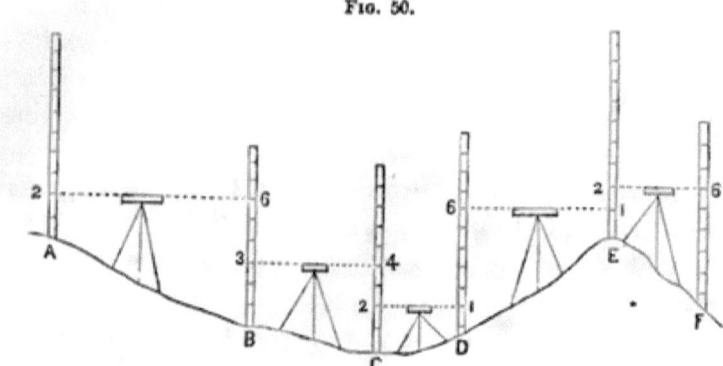

(54.) **First Form of Field-book.** In this, the names of the points, or "Stations," whose heights are demanded, are placed in the first column; and their heights, as finally ascertained, in reference to the first point, in the last column. The heights above the starting-point are marked +, and those below it are marked —. The back-sight to any station is placed on the line below the point to which it refers. When a back-sight exceeds a fore-sight, their difference is placed in the column of "Rise;" when it is less, their difference is a "Fall." The following table represents the same observations as the last figure, and their careful comparison will explain any obscurities in either:

Stations.	Distances.	Back-sights.	Fore-sights.	Rise.	Fall.	To. Heights.
A						0.00
B	100	2.00	6.00		— 4.00	— 4.00
C	60	3.00	4.00		— 1.00	— 5.00
D	40	2.00	1.00	+ 1.00		— 4.00
E	70	6.00	1.00	+ 5.00		+ 1.00
F	50	2.00	6.00		— 4.00	— 3.00
		15.00	18.00		— 3.00	

The above table shows that B is 4 feet below A; that C is 5 feet below A; that E is 1 foot above A; and so on. To test the calculations, add up the back-sights and fore-sights. The difference of the sums should equal the last "total height."

An objection to this form is that the back-sights come on the line *below* the station to which they are taken, which is embarrassing to a beginner.

When "intermediate" observations are taken, the "fore-sights," taken to these intermediate points, are put down in their proper column, and are also set down in the column of "back-sights;" so that when the two columns are added up, any error in these intermediate sights (which are usually not taken very accurately) will be cancelled, and will not affect the final result. The effect is the same as if, after the fore-sight to the intermediate point had been taken, the instrument had been taken up and set down again at precisely the same height as before, and a back-sight had then been taken to the same point. Hence, in this form, the "turning-points" are those stations which have different back-sights and fore-sights, while those which have them the same are "intermediates."

The following figure and table represent the same ground as the preceding one, but with only two settings of the instrument. D is the turning-point:

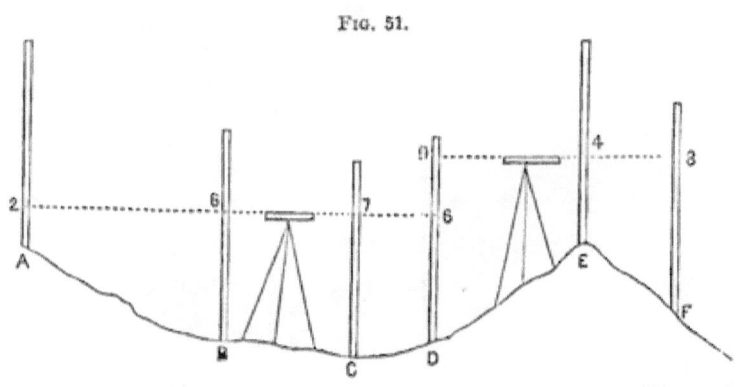

Fig. 51.

Stations.	Distances.	B. S. +	F. S. −	Rise.	Fall.	To. Heights.
A						0.00
B		2.00	6.00		4.00	− 4.00
C		6.00	7.00		1.00	− 5.00
D		7.00	6.00	1.00		− 4.00
E		9.00	4.00	5.00		+ 1.00
F		4.00	8.00		4.00	− 3.00
		+ 28.00	− 31.00		3.00	

In levelling for "sections," the distances between the points levelled must be recorded. They are usually put down after the stations *to* which they are measured; although in surveying with the compass, etc., they are put down after the stations *from* which they are measured. In the following notes, which contain intermediate stations, they are put down *before* the stations *to* which they are measured. It should be remembered that these distances are measured between the points at which the rod is held, and have no reference to the points at which the instrument is set up:

Distance.	Stations.	B. S. +	F. S. −	Rise.	Fall.	To. Heights.
	260					91.397
100	261	4.576	3.726	0.850		92.247
100	262	5.420	4.500	0.920		93.167
100	263	4.500	3.170	1.330		94.497
40	263.40	4.918	4.938		0.020	94.477
60	264	4.930	6.386		1.456	93.021
100	265	3.380	4.640		1.260	91.761
100	266	4.640	5.400		0.760	91.001
70	266.70	2.760	3.070		0.310	90.691
30	267	3.070	3.750		0.680	90.011
100	268	3.750	6.925		3.175	86.836
		41.944	46.505		− 4.561	
			41.944		+91.397	
			− 4.561		86.836	

(55.) **Second Form of Field-book.** This is presented below. It refers to the same stations and levels noted in the first table, and shown in Fig. 50:

Stations.	Distances.	Back-sights.	Ht. Inst. above Datum.	Fore-sights.	To. Heights.
A					0.00
B	100	2.00	+ 2.00	6.00	− 4.00
C	60	3.00	− 1.00	4.00	− 5.00
D	40	2.00	− 3.00	1.00	− 4.00
E	70	6.00	+ 2.00	1.00	+ 1.00
F	50	2.00	+ 3.00	6.00	− 3.00
		15.00		18.00	− 3.00

In the preceding form it will be seen that a new column is introduced, containing the Height of the Instrument (i. e., of its line of sight), not above the ground where it stands, but above the *Datum*, or starting-point, of the levels. The former columns of "Rise" and "Fall" are omitted. The preceding notes are taken thus: The height of the starting-point, or "datum," at A, is 0.00. The instrument being set up and levelled, the rod is held at A. The back-sight upon it is 2.00; therefore the height of the instrument is also 2.00. The rod is next held at B. The fore-sight to it is 6.00. That point is therefore 6.00 below the instrument, or $2.00 - 6.00 = -4.00$ below the datum. The instrument is now moved, and again set up, and the back-sight to B, being 3.00, the height of the instrument is $-4.00 + 3.00 = -1.00$, and so on; the height of the instrument being always obtained by adding the back-sight to the height of the peg on which the rod is held, and the height of the next peg being obtained by subtracting the fore-sight to the rod held on that peg, from the height of the instrument.

This form is better than the first form, in levelling for a section of the ground to make a profile; or when several observations are to be made at one setting of the level; or when points of desired heights are to be established, as in "Levelling-location," Chapter VIII.

This form may be modified by putting the back-sights on the same line with the stations to which they are taken. This avoids the defect of the first form, but introduces the new defect of writing them down after the number which they precede, in a back-handed way, which may be a source of error.

This modification is shown in the following table, which corresponds to Fig. 51. In the column of fore-sights, the "turning-points" (T. P.), and "intermediate points" (Int.), are put in separate columns; so that, to prove the work, the difference of the sum of the back-sights and of the sum of the turning-point fore-sights, is the number which should equal the difference of the heights of the first and last points:

Stations.	Distances.	B. S. +	Ht. of Inst.	F. S. −		To. Heights.
				T. P.	Int.	
A		2.00				0.00
B			+ 2.00		6.00	− 4.00
C					7.00	− 5.00
D		9.00		6.00		− 4.00
E			+ 5.00		4.00	+ 1.00
F				8.00		− 3.00
		+ 11.00		− 14.00		
				+ 11.00		
				− 3.00		

When a line is divided up into stations of 100 feet each, as on railroad work, the number of the station indicates its distance from the starting-point. When an observation is taken at a point between these hundred-feet stations, it is noted as a decimal thus: Station 4.60 is 460 feet from the starting-point. In the field-notes of such work, the column of distances may be omitted, as in the following table. The heights and distances are the same as in the last table under Art. (54):

Stations.	B. S. +	Ht. of Inst.	F. S. −		Total Heights.
			T. P.	Int.	
260	4.576	95.973			91.397
261	5.420	97.667	3.726		92.247
262				4.500	93.167
263	4.910	99.407	3.170		94.497
263.40				4.938	94.477
264	3.380	96.401	6.386		93.021
265				4.640	91.761
266	2.760	93.761	5.400		91.001
266.70				3.070	90.691
267				3.750	90.011
268			6.925		86.836
	+ 21.046		− 25.607		
			+ 21.046		
			− 4.561		
			+ 91.397		
			+ 86.836		

(56.) **Third Form of Field-book.** In this form, the defects of the preceding methods are avoided, and it approximates to a sketch of the operations; the fore-sights being placed before the stations to which they are taken, and the back-sights after them. The distances are placed before the stations to which they are taken; or after those from which they are taken. Another advantage is that the stations, their heights, and the distances, are brought together; which facilitates the making of a profile. The following table is the case given in Fig. 50:

F. S. −	Distances.	Stations.	Ht. of Peg.	B. S. +	Ht. of Inst.
		A	0.00	2.00	+ 2.00
6.00	100	B	− 4.00	3.00	− 1.00
4.00	60	C	− 5.00	2.00	− 3.00
1.00	40	D	− 4.00	6.00	+ 2.00
1.00	70	E	+ 1.00	2.00	+ 3.00
6.00	50	F	− 3.00		
− 18.00				+ 15.00	
				− 18.00	
				− 3.00	

When "intermediates" are taken, the first column may be divided into two heads (as in the second table, Art. 55), respectively "turning-points" (T. P.), and "intermediate points" (Int.). The work is tested by taking the difference of the sum of the "T. P.'s" and "B. S.'s" The symbol ⊖ is used to represent the height of the cross-hairs. This table is for Fig. 51:

F. S. −		Stations.	Distances.	Ht. of Peg.	B. S. +	⊖
T. P.	Int.					
		A	100	0.00	2.00	+ 2.00
	6.00	B	60	− 4.00		
	7.00	C	40	− 5.00		
6.00		D	70	− 4.00	9.00	+ 5.00
	4.00	E	50	+ 1.00		
8.00		F		− 3.00		
− 14.00					+ 11.00	
					− 14.00	
					− 3.00	

Fourth Form of Field-book. In this the back sights are placed directly under the height of the station to which they are taken, which lessens the chance of making mistakes in adding to get the height of instrument. The height of instrument is distinguished by being included between two horizontal lines. The following table refers to Fig. 51.

Station.	F. S.	Heights.	Remarks.
A		0.00	
		2.00	
		2.00	
B	6.00	—4.00	
C	7.00	—5.00	
D	6.00	—4.00	
		9.00	
		5.00	
E	4.00	1.00	
F	8.00	—3.00	

(57.) **Best Length of Sights.** There are two classes of inaccuracies. With very long sights, the errors of imperfect adjustment and curvature are greatest; the former varying as the length, and the latter as the square of the length. With very short sights, and therefore more numerous, the errors of inaccurate sighting at the target are greatest. The best usual mean is from 200 feet to 300 feet, or more if equal distances for back-sights and fore-sights to turning-points can be obtained.

(58.) **Equal Distances of Sight.** They are always very desirable. They are most easily determined, when no stakes have been previously set, by "stadia" cross-hairs in the telescope of the level. [L. S., 375.]

(59.) **Datum-Level.** This is the plane of reference, from which, above it or below it, usually the former, the heights of all points of the line are reckoned.

It may be taken as the height of the starting-point. If the line descends, it is better to call the starting-point 10 feet or 100 feet above some imaginary plane, so that points below the starting-point may not have minus signs.

It is desirable to refer all levels in a country to some one datum. This is usually the surface of the sea, and for general purposes *mean tide* is best. *Low-water* mark should be the datum when the levellings are connected with harbor-surveys, whose soundings always refer to low water. *High-water* mark should be used when the levellings relate to the drainage of a country.

(60.) **Bench-Marks (B. M.).** These are permanent objects, natural or artificial, whose heights above the datum are determined and recorded for future reference.

Good objects are these: pointed tops of rocks; tops of milestones; stone door-sills; tops of gate-posts or hinges; and generally any object not easily disturbed, and easily described and found.

A knob made on the spreading root of a tree is good. A nail may be driven in it, and the tree "blazed" and marked, as in Fig. 52. A stake will do till frost.

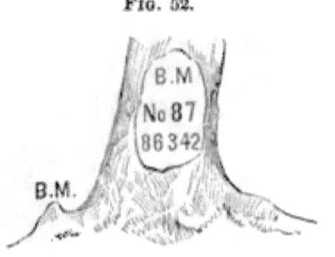
Fig. 52.

Bench-marks should be made near the starting-point of a line of levels; near where the line crosses a road; on each side of a river crossed by it; at the top and bottom of any high hill passed over; and *always* at every half-mile or mile.

The precise location and description of every B. M. should be noted very fully and precisely, and in such a way that an entire stranger could find it, with the aid of the notes.

(61.) **Check-Levels, or Test-Levels.** No *single* set of levels is to be trusted; but they must be tested by another set, run between the bench-marks (B. M.'s), though not necessarily over the same ground.

A set of levels will verify themselves if they come around to the starting-point again.

(62.) Limits of Precision. Errors and inaccuracies should be carefully distinguished. For the latter, every leveller must make a standard for himself, so as to be able, in testing his work, to distinguish any *real error* from his *usual inaccuracy*.

The result of four sets of levellings, in France, of from 45 to 140 miles, averaged a difference of $\frac{1}{10}$ ft. in 43 miles, and the greatest error was $\frac{1}{8}$ ft. in 56 miles.

A French leveller, M. Bourdaloue, contracts to level the B. M.'s of a R. R. survey to within 0.002 ft. per mile, or $\frac{1}{10}$ ft. per 50 miles.

In Scotland, the difference of two sets of levels of 26 miles was 0.02 ft.

(63.) Trial-Levels, or Flying-Levels. Their object is to get a general approximate idea of the comparative heights of a portion of the country, as a guide in choosing lines to be levelled more accurately. More rapidity is required, and less precision is necessary. The distances may be measured at the same time by stadia-hairs.

(64.) Levelling for Sections. The object of this is to measure all the ascents and descents of the line, and the distances between the points at which the slope changes; so that a section or profile of it can be made from the observations taken.

The line of a railroad is usually set out by a party with compass or transit, who drive at every hundred feet a large stake with the number of the station on it, and beside it a small level-peg, even with the surface of the ground. On this the rod is held for the observations. The level-peg is set in "line," and the large stake a foot or two to one side.

(65.) Profiles. A profile is a section of ground by a vertical plane or cylindrical surface,[1] passing through the line along

---

[1] A cylindrical surface is here understood to mean that formed by a line moving parallel to itself along *any* line, instead of only a circle, as in elementary geometry.

which a profile is desired. It represents to any desired scale the heights and distances of the various points of a line, its ascents and descents, as seen in a side view. It is made thus: Any point on the paper being assumed for the first station, a horizontal line is drawn through it; the distance to the next station is measured along it, to the required scale; at the termination of this distance a vertical line is drawn; and the given height of the second station above or below the first is set off on this vertical line. The point thus fixed determines the second station, and a line joining it to the first station represents the slope of the ground between the two. The process is repeated for the next station, etc.

But the rises and falls of a line are always very small in proportion to the distances passed over, even mountains being merely as the roughnesses of the rind of an orange. If the distances and the heights were represented on a profile to the same scale, the latter would be hardly visible. To make them more apparent, it is usual to "exaggerate the vertical scale" tenfold, or more; i. e., to make the representation of a foot of height ten times as great as that of a foot of length, as in Fig. 50, in which one inch represents one hundred feet for the distances, and ten feet for the heights.

In practice, engraved profile-paper is generally used, which is ruled in squares or rectangles, to which any arbitrary values may be assigned.

When the line levelled over is not straight, the profile, whose length is that of the line straightened out, will extend beyond the "plan" when both are on the same sheet.

(66.) **Cross-Levels.** These show the heights of the ground on a line at right angles to the main line. They give "cross-sections" of it. In the note-book they are put on the right-hand page. They may be taken at the same time with the other levels, or independently. In taking cross-levels where the slopes are quite steep, as in mountain districts, frequent settings of the instrument are necessary.

A much more rapid method is by the use of "cross-section rods." These are two rods, one of which is about ten or twelve feet long, provided with a bubble-tube near each end, so as to be held level, and graduated to feet, tenths, and hundredths. The other is simply a graduated rod. The manner of using them is shown in Fig. 53.

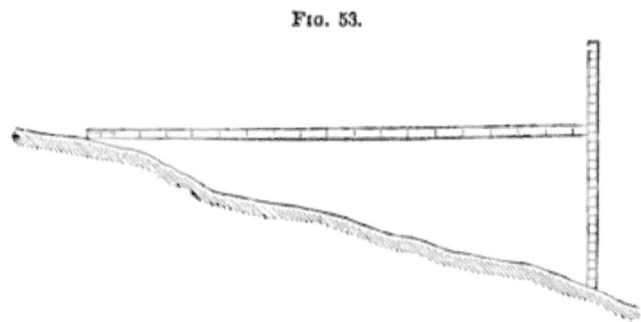

Fig. 53.

A slope-level is sometimes used. See "Angular Surveying," Part II.

CHAPTER VII.

DIFFICULTIES.

(67.) Steep Slopes. In descending or ascending a hill, the instrument and the rod should be so placed that the sight should strike as near as possible to the bottom of the rod on the up-hill side, and the top of the rod on the down-hill side.

Try this by levelling over two screws, setting the instrument so that one pair of opposite plate-screws shall point in the direction of the line, but do not be too particular; it is a waste of time.

Doing this produces sights of unequal length. The rod being about twice as high as the instrument, the down-hill sights will be about double the length of the up-hill ones, as shown in Fig. 54. Then set to one side of the line. This is

FIG. 54.

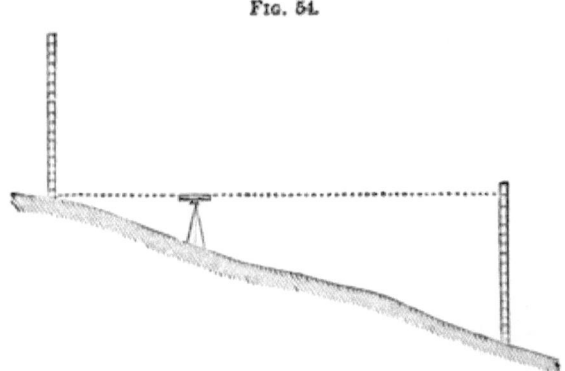

necessary on slopes so steep that the rod is too near the level to be read. If this be impossible, keep notes of the lengths of the sights to the turning-points, backward and forward, and as soon as possible take sights unequal in the contrary direction till the differences of lengths balance the former ones. When approaching a long ascent or descent, make these compensations in advance.

In levelling over a line of stakes already set, as on a railroad, at every 100 ft., if the line of sight strikes not quite up to one, drive a peg as high as you can see it, and make it a turning-point, noting it "peg" in the field-book.

In levelling across a hill or hollow, instead of setting the instrument on the top of the hill or bottom of the hollow, time will be saved by the method represented in Figs. 55 and 56.

FIG. 55.

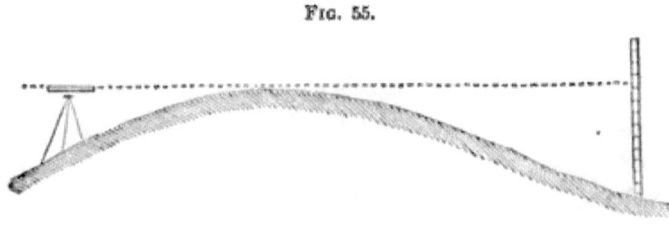

(68.) *When the rod is a little too low,* raise it alongside of a stake, or the body, and put the top of the rod "right;" then measure down from the bottom of the rod, and add it to its length.

DIFFICULTIES.

Fig. 56.

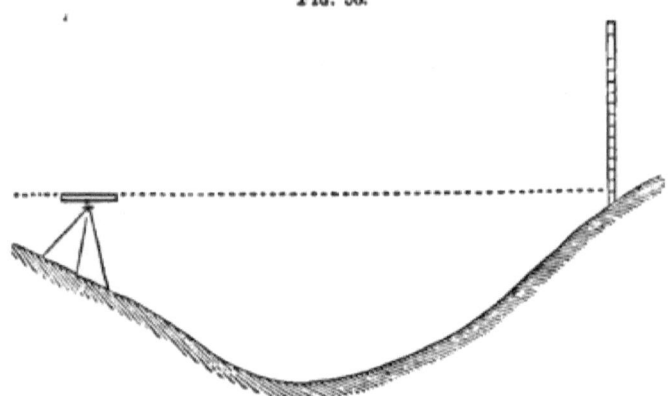

(69.) *When the rod is a little too high,* so that the line of sight strikes the peg below the bottom of the rod, measure down from the top of the peg, and put down the sight with a contrary sign to what it would have had; i. e., if a back-sight make it minus, and if a fore-sight make it plus.

(70.) *When the rod is too near.* When no figure is visible, raise the rod slowly till a figure comes in sight. If too near to read, and there is no target, use a field-book as target. If the instrument is exactly over the peg, measure up to the height of the cross-hairs, as given by the side-screws.

(71.) WATER. A.—*A pond too wide to be sighted across.* Drive a peg to the level of the water, on the first side, and observe its height, as an F. S. Then drive a peg on the other side of the pond, also to the surface of the water. Hold the rod on it. Set up the level beyond it, and sight to it as a B. S., and put down the observation as if it had been taken to the first peg.

Fig. 57.

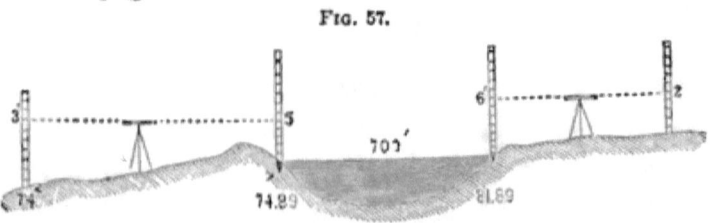

F. S.	Sta.	Ht.	B. S.	⊙
	74	50.00	3.00	53.00
5.0	74.89 }	48.00		
	81.89 }		6.00	54.00

There must be no wind in the direction of the line of level.

B.—*For levelling across a running stream.* Set the two pegs in a line at right angles to the current, although the line to be levelled may cross it obliquely.

If a profile or section of the ground under the water be required, find the height of the surface, and measure the depths below this at a sufficient number of points, measuring the distances also, and put these depths down as fore-sights.

(72.) **A Swamp, or Marsh.** This cannot be treated like a pond, for the water may seem nearly stagnant while its surface has considerable slope, its flow being retarded by vegetation. If only slightly "shaky," have an observer at each end of the level. If more so, push the legs down as far as they will go, and let both observers lie down on their sides. If still more "shaky," drive three stakes or piles, to support the legs of the tripod, and stand the tripod on them.

A water-level will level itself. Use that for intermediate points on the swamp, and test the result by levelling *around* the swamp with the spirit-level.

(73.) **Underwood.** If it cannot be cut away, set the instrument on some eminence, natural or artificial.

(74.) **Board Fence.** Run a knife-blade through one of the boards, and hold the rod upon it on each side of the fence, as if it were a peg, keeping the blade in the same horizontal position while the rod and instrument are taken over.

(75.) **A Wall.** *First Method.* Drive a peg at the bottom of the wall, on the first side, and observe on it. Measure the height of the wall above the peg, and put this down as a B. S. Drive another peg on the other side of the wall; measure down to it from the top of the wall, and put that down as an F. S., just as if the level had been set in the air at the height of the top of the wall, and this B. S. and F. S. had been really taken. Set up the instrument beyond the wall, take a B. S. to this peg, and go on as usual.

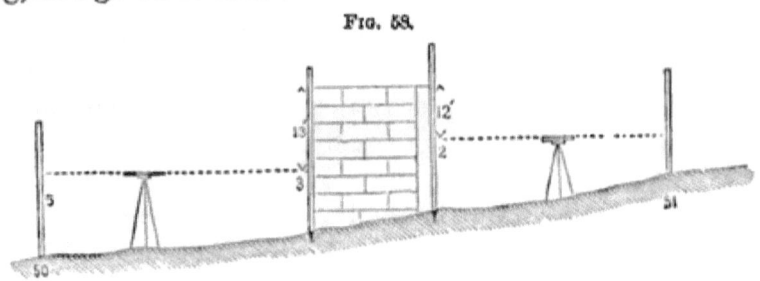

Fig. 58.

F. S.	Sta.	Ht.	B. S.	⊖
	50	74.00	5.00	79.00
3.00	Peg.	76.00	13.00	89.00
12.00	Peg.	77.00	2.00	79.00
1.00	51	78.00		

Second Method. Mark where the line of sight strikes the wall; measure up to the top of the wall, and put this down as an F. S., with a plus sign, as in (69), where the line of sight struck below the top of the peg.

On the other side of the wall, sight back to it, and mark where the line of sight strikes. Measure to the top of the wall, and put this down as a B. S., with a minus sign, and then go on as usual.

(76.) **House.** First try to find some place for the instrument from which you can see through, by opening doors or windows. Or, find some place in the house where you can set the instrument and see both ways, or hold the rod at some point inside, and look to it from front and back. A straight stick may be used if the rod cannot be held upright, and the height measured on the rod.

(77.) **The Sun.** It often causes the leveller much difficulty.
1. By shining in the object-glass. If the instrument has a

shade on it, draw it out. If not, shade the glass with your hand or hat, or set the instrument to one side of the line.

2. By heating the level unequally in all its parts. Holding an umbrella over it will remedy this.

3. By causing irregular refraction. Some parts of the ground become heated more than others, and therefore rarefy the air at those places. This cannot be avoided nor corrected.

(78.) **Wind.** Watch for lulls of wind, and observe then several times, and take the mean. The least wind is at daybreak.

(79.) **Idiosyncrasies:** Different persons do not see things precisely alike. Each individual may have an inaccuracy peculiar to himself. One may read an observation higher or lower than another equal in skill. Also, a person's right and left eye may differ. This difference in individuals is termed their "personal equation."

To test the accuracy of your eye, turn the head so as to bring the eyes in the same vertical line, and sight to the rod held horizontally. Note where the vertical hair strikes. Then turn the head to the other side, so as to invert the position of the eyes, and then sight again. As before, the mean of the two readings is the correct one.

(80.) **Reciprocal Levelling.** This is to be used when it is impossible to set midway between the two points.

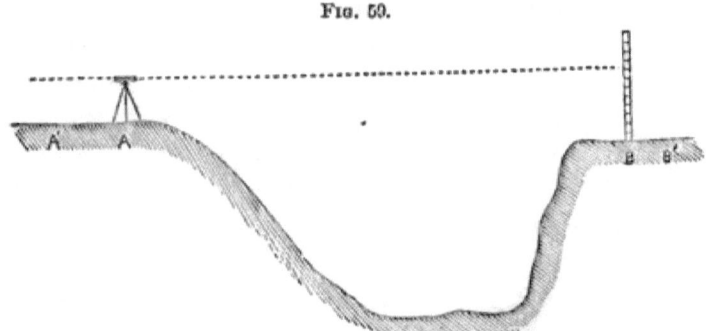

Fig. 59.

Set the instrument over A, and sight to rod at B, and note reading. The difference of the reading and of the height of the cross-hairs gives *a* difference of height of A and B. Then set up at B, and observe to A, similarly. A new difference of height is obtained. The mean of these two is the correct one.

Ht. of cross-hairs above peg at A $= 4'.3$ Ht. of cross-hairs above peg at B $= 4'.9$
            Observation to B $= 7'.0$ Observation to A $= 4'.2$
            Diff. of height $= 2'.7$ Diff. of height $= 0'.7$
                     True difference $= \frac{1}{2}(2'.7 + 0'.7) = 1'.7$.

Otherwise, set the instrument at an equal distance from each point, as A′ and B′, and observe to each in turn. The mean of the two differences of height obtained will be the true difference, as before.

CHAPTER VIII.

LEVELLING LOCATION.

(81.) Its Nature. It is the converse of the general problem of levelling, which is to find the difference of heights of two given points. *This* consists in determining the place of a point of any *required* height above or below any given point.

To do this, hold the rod on some point of known height above the datum-level; sight to it, and thus determine the height of the cross-hairs. Subtract from this the desired height of the required point, and set the target at the difference. Hold the rod at the place where the height is desired, and raise or lower it till the cross-hair bisects the target. Then the bottom of the rod is at the desired height. Usually, a peg is driven till its top is at the given height above the datum.

(82.) Difficulties. If the difference of height be too much to be measured at one setting of the instrument, take a series of levels up or down to the desired point. So, too, if they be far apart; and thus find a place where, the instrument having a known height of cross-hairs, the target can finally be set, as before.

If the ground be so low or so high that a peg cannot be set with its top at the required height, drive a peg till its top is just above the surface of the ground. Observe to the rod on it, determine its height above or below the desired point, and note this on a large stake driven beside it; or, place its top a whole number of feet above or below the required height, and mark the difference on it, or on a stake beside it.

(83.) **Staking out Work.** When embankments and excavations are to be made for roads, etc., side-stakes are set at points in their intended outside edges; i. e., where their slopes will meet the surface of the ground; and the height which the ground at those points is above or below the required height or depth of the top or bottom of the finished work, is marked on these stakes with the words "cut," or "fill," or the signs + or −.

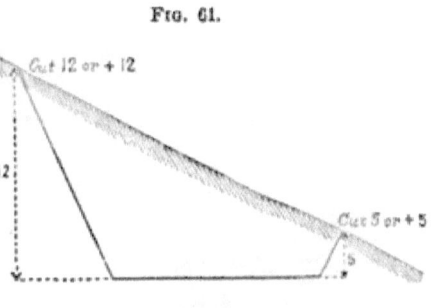

The places of the stakes are found by trial. (See Gillespie's Road-making, p. 145.) These stakes are set to prepare the work for contractors. When the work is nearly finished, other stakes are set at the exact required height.

In staking out *foundation-pits*, set temporary stakes exactly above the intended bottom angles of the completed pit, thus marking out on the surface of the ground its intended shape. Take the heights of each of these stakes and move them outward such distances that cutting down from them with the proper depth and slope will bring you to the desired bottom angle.

(84.) To locate a Level-Line. This consists in determining on the surface of the ground a series of points which are at the same level; i. e., at the same height above some datum. Set one peg at the desired height, as in (81). Sight to the rod held thereon, and make fast the target when bisected. Then send on the rod in the desired direction, and have it moved up or down along the slope of the ground, until the target is again bisected. This gives a second point. So go on as far as sights can be correctly taken, keeping unchanged the instrument and target. Make the last point sighted to a "turning-point." Carry the instrument beyond it, set up again, take a B. S., and proceed as at first.

The rod should be held and pegs driven at points so near together that the level-line between them will be approximately straight.

(85.) Applications. One use of this operation is to mark out the line which will be the edge of the water of a pond to be formed by a dam. In that case, a point of a height equal to that of the top of the proposed dam, *plus* the height which the water will stand on it (to be determined by hydraulic formulas), will be the starting-point. Then proceed to set stakes as directed in the last article.

The line from stake to stake may then be surveyed like the sides of a field, and the area to be overflowed thus determined.

Strictly, the surface of the water behind a dam is not level, but is curved concavely upward, and is therefore higher and sets back farther than if level. This backing up of the water is called *Remous*.

Another important application of this problem is to obtain "contour lines" for Topography.

(86.) **To run a Grade-Line.** This consists in setting a series of pegs so that their tops shall be points in a line which shall have any required slope, ascending or descending.

When a grade-line is to be run straight between two given points, set the level over one point, set the target at the height of the cross-hairs, hold the rod on the other point, and raise or lower one end of the instrument till the cross-hair bisects the target. Then send the rod along the line, and drive pegs to such heights that when the rod is held on them the cross-hair will bisect the target. A stake may be driven at the extreme point to the height of the target.

A line of uniform grade or slope is not a straight line. Calling the globe spherical, this line, when traced in the plane of a great circle, would be a logarithmic spiral. On a length of six miles, the difference in the middle between it and its straight chord would be six feet.

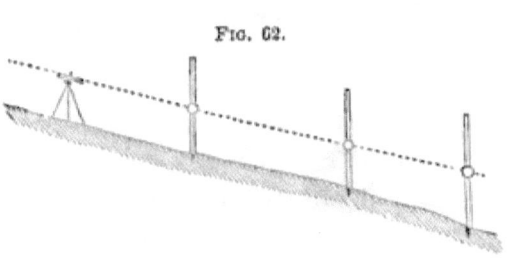

FIG. 62.

PART II.

INDIRECT LEVELLING.

CHAPTER I.

METHODS AND INSTRUMENTS.

(87.) Vertical Surveying. Levelling may be named VERTICAL SURVEYING, or *Up-and-down Surveying;* Land Surveying being HORIZONTAL SURVEYING, or *Right-and-left* and *Fore-and-aft Surveying.*

All the methods of determining the position of a point in horizontal surveying, may be used in vertical surveying.

The point may be determined by coördinates situated in a vertical plane, as in any of the systems employed in L. S. (Part I., Chapter I.), in a horizontal plane.

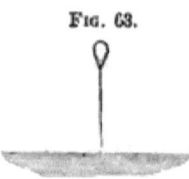

FIG. 63.

Thus, if a balloon be held down by a single rope attached to a point in a level surface, its height above that surface is found by measuring the length of the rope. This is the Direct Method. It resembles that of "rectangular coördinates," L. S. (6); though here only one of the coördinates is measured. The other might be situated anywhere in the surface.

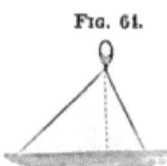

FIG. 64.

If, however, the balloon be held down by two cords, its height can be determined by measuring the length of the cords and the distance between their lower ends. They correspond to the "focal coördinates" of L. S. (5). The required vertical height can be calculated by trigonometry. So in the following other Indirect Methods.

The length of the string of a kite, and the angle which this string makes with the horizon, are the "polar coördinates" of the kite; as in L. S. (7).

The "angles of elevation" of a meteor, observed by two persons in the same vertical plane with it, and at known distances apart, are its "angular coördinates," as in L. S. (8).

Finally, an aëronaut could determine his own height by observing the angles subtended by three given objects situated on the earth's surface, at known distances, and in the same vertical plane with him. These angles would be the "trilinear coördinates" of L. S. (10).

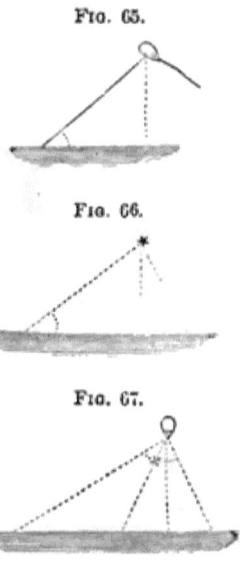

FIG. 65.

FIG. 66.

FIG. 67.

Many other systems of coördinate lines and angles, variously combined, may be employed.

The desired heights may also be determined by various other methods, analogous to those given in L. S. for "inaccessible distances."

Combinations of measurements not in the same vertical plane may also be used, as will be shown in Chapter III.

(88.) Vertical Angles. The vertical angles measured may be those made—either with a level line, or with a vertical line—by the line passing from one point to the other.

The angle BAC is called an "angle of elevation," and the angle $B'AC$ an "angle of depression." The former angle may be called positive, and the latter negative.

The angle BAZ or $B'AZ$ is

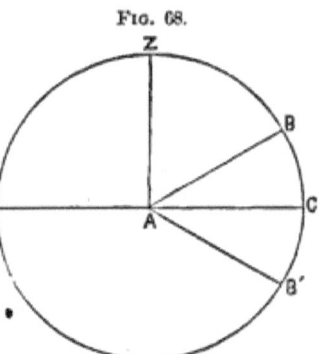

FIG. 68.

called the zenith distance of the object. It is the complement of the former angle, i. e., $= 90° -$ that angle taken with its proper algebraic sign. An angle of elevation, $B A C = 10°$, would be a zenith distance of 80°. An angle of depression, $B'A C = - 10°$, would be a zenith distance of 100°. The zenith distance is preferable in important and complicated operations, as avoiding the ambiguity of the other mode of notation.

(89.) **Instruments.** All contain a divided circle, or arc, placed vertically, and a level or plumb line. By these is measured the desired vertical angle made by the inclined line with either a level or vertical line.

This inclined line may be an actual line or a visual line. In the former case, it may be a rod, or cord, or wire, as shown in the figures:

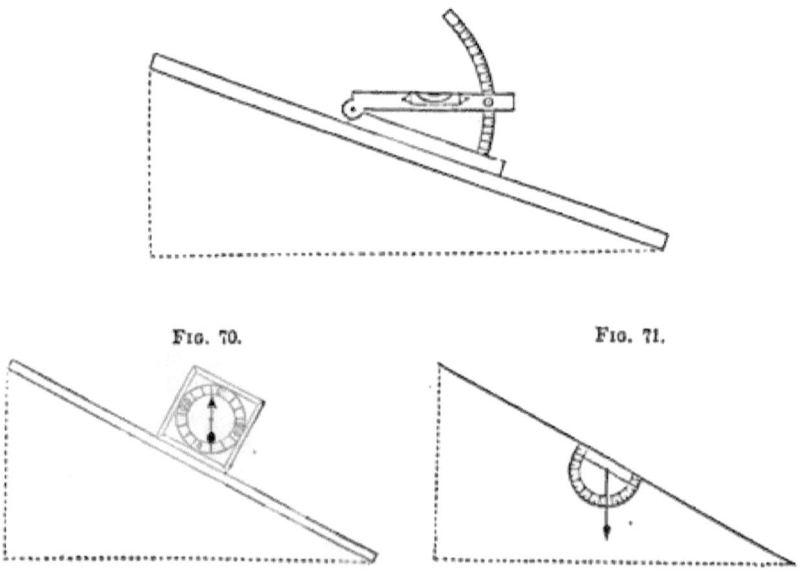

FIG. 69.

FIG. 70.   FIG. 71.

This last arrangement of a cord or wire, Fig. 71, is used in mine surveying. A light surveyor's chain may be similarly used, with the advantage of giving, at the same time, difference of heights and distance.

Diff. of hts. = length of chain × sin. angle.
Hor. distance = length of chain × cos. angle.

These instruments are all "Slope-measurers." They are also called *Clinometers, Clisimeters, Eclimeters,* etc., all meaning the same thing.

(90.) **Slopes.** These may be designated by their angles with the horizon, or by the relations of their bases and heights. The French engineers name a slope by the ratio of its height to its base; i. e., $\frac{BC}{AC}$; which is the tangent of the angle BAC. The English and Americans use the ratio of the base to the height; i. e., $\frac{AC}{BC}$, and make the height the unit, so that if $AC = 2\,CB$, the slope is called 2 to 1; and so on.

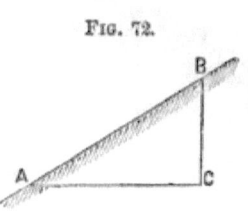

Fig. 72.

(91.) When the inclined line is a visual line, such as the line of sight of a telescope, whose angular movements are measured on a vertical circle beside it, and when with these is combined a horizontal circle for measuring horizontal angles, the instrument is called a "Theodolite."

In the usual American form, the telescope turns over. It is a transit-theodolite. (See Fig. 73.) It is usually called simply a "Transit."

For the usual English form, see L. S., page 213.

In the usual French form, the telescope is eccentric; i. e., on one side of the vertical axis, and has a counterpoise on the other side, as in Fig. 141, of mining transit.

(92.) **The Surveyor's Transit,** Fig. 73. The telescope revolves on a horizontal axis, which itself rests on two standards, S S, attached to the horizontal vernier-plate, H. The graduated vertical circle, A, by which vertical angles are measured, is attached to the telescope axis, and is read with a vernier on the lower side. A level, L, is attached to the telescope, in the same manner as that of the Y level. The vernier-plate, which carries the telescope, is furnished with two verniers on opposite sides of the instrument, and at right angles to the telescope. The vertical and the horizontal graduated circles are both furnished with a clamp and slow-

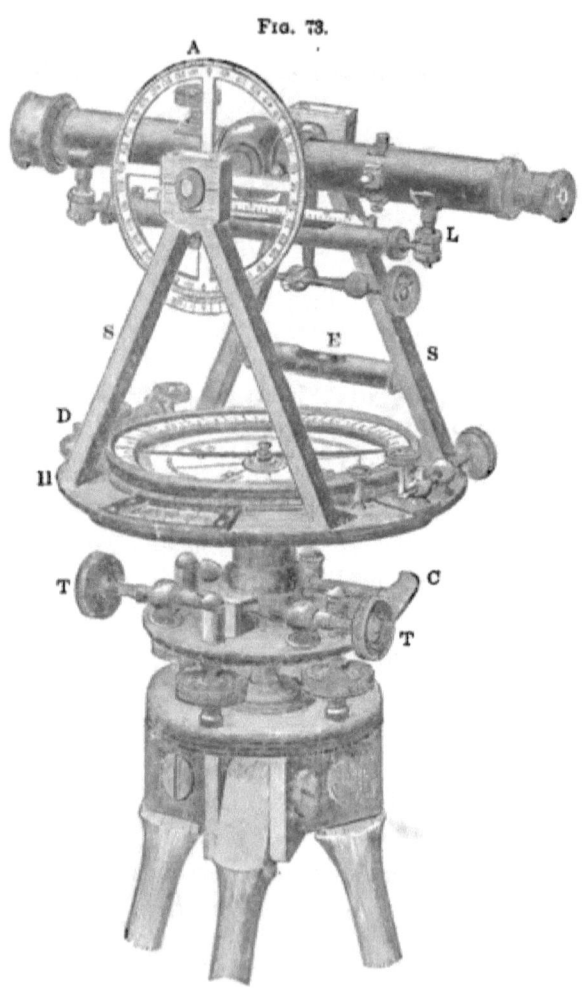

Fig. 73.

motion screw. Attached to the upper parallel plate is another clamp, C, and a pair of slow-motion screws, T T, by which all of the instrument above the clamp may be given a slow motion, horizontally. The vernier-plate is furnished with two levels, at right angles to each other. One of them, D, is attached to the plate, and the other, E, is fastened to the standard, up out of the way of the second vernier.

The compass may be used like a common surveyor's compass, the telescope taking the place of the sights. Its principal use is to serve as a check on the observations, the difference of the magnetic bearings of two lines being approximately equal to the angle measured between them by the more perfect instrument.

The arrangement of the parts of the telescope, and the parallel plates, are the same as for the Y level.

(93.) **Adjustments.** *First Adjustment.* To cause the bubbles to remain in the centre of the tubes, when the vernier-plate is turned around horizontally; i. e., to make the plane of the levels perpendicular to the vertical axis of the instrument:

To test this, turn the vernier-plate till each of the plate-levels is parallel to an opposite pair of the parallel plate-screws, and bring each bubble to the middle of its tube, by the screws to which it is parallel. Then turn the plate half-way around. If either of the bubbles runs from the centre of the tube, bring it half-way back, by raising or lowering one end of the tube, and the rest of the way, by the parallel plate-screws. Again, turn the plate half-way around, and repeat the operation, if necessary. The other tube must be tested, and, if necessary, adjusted in the same way.

Second Adjustment. To cause the line of collimation to revolve in a plane; i. e., to make the line of collimation perpendicular to its axis:

Set up the instrument and level it carefully. Sight to some well-defined point, as far off as can be distinctly seen.

FIG. 74.

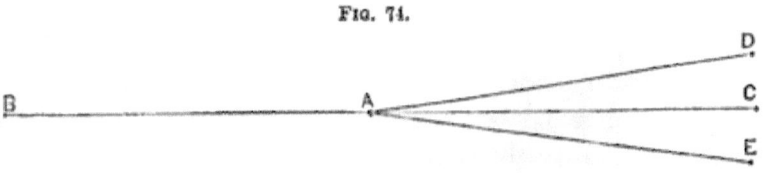

Clamp the instrument so that there can be no movement horizontally, turn the telescope over, and fix another point (as a nail driven in a stake) precisely in the line of sight, and at an equal distance from the instrument.

In the figure let A be the place of the instrument and B the first point sighted to. If the vertical cross-hair is in adjustment, the line of sight, on turning over the telescope, will strike at C, A C being a prolongation of the straight line A B. If not in adjustment, it will strike on one side, as at D. Now loosen the clamp, turn the vernier-plate half-way around, and sight to the first object selected. Again clamp the instrument and turn over the telescope. The line of sight will now strike at E, as far to the right of the true line as D is to the left.

To correct this, move the vertical cross-hair till the line of sight strikes half-way between E and C. Verify again, and repeat the operation, if necessary.

Third Adjustment. To cause the line of collimation to move in a truly vertical plane when the telescope is revolved; i. e., to make the axis of the line of collimation parallel to the plane of the levels:

Set up the instrument near the base of a spire, or other high object, and level it carefully. Sight to some well-defined point on the top of the object. Clamp the instrument so that there can be no motion horizontally, turn down the telescope and fix a point at the base of the object, precisely in the line of sight. Now loosen the clamp, turn the vernier-plate half-way around, and sight again to the point on the top of the object. Again clamp the instrument, and turn down the telescope. If in adjustment, the line of sight will again strike the point fixed at the base. If not, the apparent error is double the real error. Make the adjustment by raising or lowering one end of the axis by means of a screw, placed in the standard for that purpose.

Fourth Adjustment. To cause the line of collimation of the telescope to be horizontal when the bubble of the level attached to it is in the centre of its tube:

The verification and adjustment is the same as the operation of adjusting the Y level by the " peg method," Art. (38).

The operations of centring the object-glass and eye-piece are the same as for the level, Art. (37).

Another adjustment is necessary in order that the vernier of the vertical circle may read zero when the bubble is in the centre. This is verified in various ways:

1. *By simple inspection.*
2. *By reversion.* Sight to some point. Note the reading on the vertical circle. Turn the telescope half-way around horizontally. Turn over the telescope and again observe the same point, and note the reading. Half the difference (if any) of the two readings is the error.

The principle is that given in L. S. (334).

This method requires the instrument to be a transit-theodolite.

3. *By reciprocal observations.* Observe successively from each of two points to the other. Half the difference of the readings equals the index-error.

When the verification has been made, the error may be rectified on the instrument, or noted as a correction to each observation, when the instrument is large and delicate.

(94.) **Field-Work.** *To measure horizontal angles.* Set the transit so that its centre shall be precisely over the angular point. This is done by means of a plumb-line, suspended from the centre of the instrument. Level the instrument carefully. Sight to a rod, held at some point on one of the lines, as at B in the figure (A being the place of the transit), and note the reading. Then loosen the clamp of the vernier-plate, keeping the other plate clamped; sight to a rod held at some point on the second line, as at C, and again note the

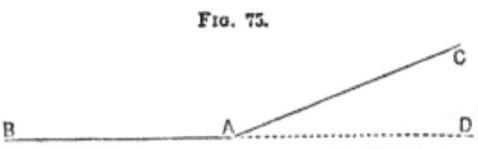

Fig. 75.

reading. The difference of the two readings will give the angle B A C. This is the *angle of intersection.*

To measure the *angle of deflection*, D A C, i. e., the angle between A C and B A prolonged: After sighting to B, turn over the telescope. It will now point toward D, in the line B A prolonged. Note the reading, sight to C, and again note the reading. The difference of the readings will give the required angle.

Vertical angles are measured similarly to horizontal ones, only using the vertical instead of the horizontal circle.

Traversing. In this method of surveying and recording a line, the direction of each successive portion is determined, not by the angle which it makes with the line preceding it, but with the first line observed, or some other constant line. The operation consists essentially in taking each back-sight by the lower motion (which turns the circle without changing the reading), and taking each forward sight by the upper motion, which moves the vernier over the arc measuring the new angle; and thus adds it to or subtracts it from the previous reading.

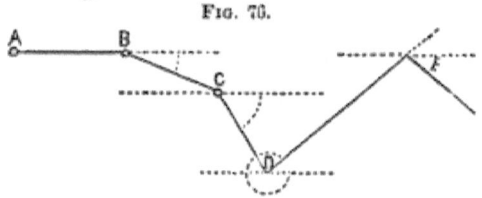

Fig. 76.

Set up the instrument at some station, as B; put the vernier at zero, and, by the lower motion, sight back to A. Tighten the lower clamp, reverse the telescope, loosen the upper clamp, sight to C by the upper motion, and clamp the vernier-plate again. Remove the instrument to C, sight back to B by the lower motion. Then clamp below, reverse the telescope, loosen the upper clamp, and sight to D by the upper motion. Then go to D and proceed as at C; and so on. The reading gives the angles measured to the right or "with the sun," as shown by the arcs in the figure.

(95.) Angular Profiles. A section or profile of a tolerably uniform slope is most easily obtained, as shown in the figures, by measuring the heights or depths below an inclined line, instead of below a level line.

Fig. 77.

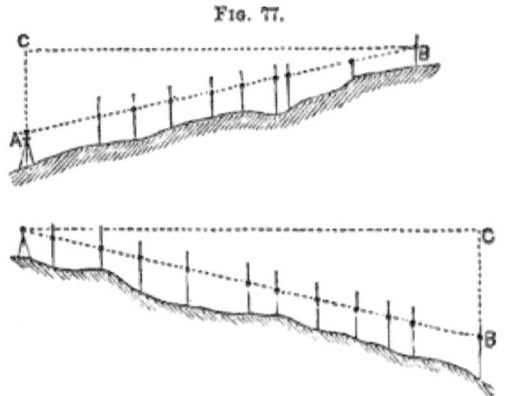

A cross-section for a road may be taken in the same way.

(96.) **Burnier's Level.** It is a pear-shaped instrument, having two graduated circles; one vertical, having a weight attached so as to keep it in the same vertical position when in use; and the other, a horizontal graduated circle, made light and carried around by a magnetic needle, so that the instrument can be used as a compass as well as a slope or angular level. It has a convex-glass, or lens, in the smaller end, through which can be seen a hair which covers, on the circle, the number of the degrees of the angle of inclination, or of the horizontal angle.

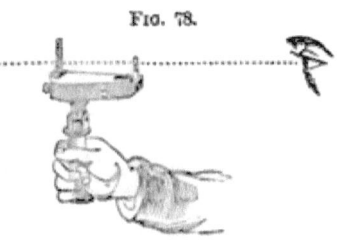

Fig. 78.

The sights are on the top or sides, according as it is used as a compass or slope-measurer. It is used by sighting to the object, and at the same time reading off the angle, the hair covering the zero-mark when the instrument is level.

(97.) **German Universal Instrument.** Its use is to enable the observer to sight to an object nearly or quite overhead. It consists of a telescope having the part which carries the eye-piece at right angles to the part carrying the object-glass, instead of being in the same straight line, as in an

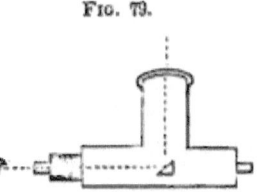

Fig. 79.

ordinary telescope. The part containing the eye-piece is connected with the other part at the axis, and is in the same line with the axis.

In the telescope is placed a small mirror, or reflector, or (what is still better) a triangular prism of glass, at an angle of 45° to the line of sight. Thus the observer can keep his eye at the same place at any inclination of the telescope.

CHAPTER II.

SIMPLE ANGULAR LEVELLING.

A.—*For Short Distances.*

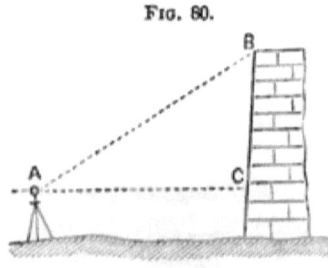

Fig. 60.

(98.) **Principle.** For short distances, curvature and refraction may be neglected. Thus, if the height of a wall, house, tree, etc., be desired, note the point where the horizontal line strikes the wall, etc., and add its height above the ground to that calculated by the formula:

$$BC = AC . \tan BAC. \qquad [1.]$$

(99.) The "best-condition" angle for observation (see L. S., 383) is 45°. Hence, in setting the instrument, we should, where practicable, have the distance about equal to the height of the point whose height we wish to ascertain.

B.—For Greater Distances.

(100.) Correction for Curvature. A C is the line of apparent level, as given by the instrument, and A C′ is the line of true level. Calling the angle A C B = 90° (which it is approximately for moderately great distances), formula [1] gives B C as the height of B above A. But B C′ is the true difference of heights of A and B.

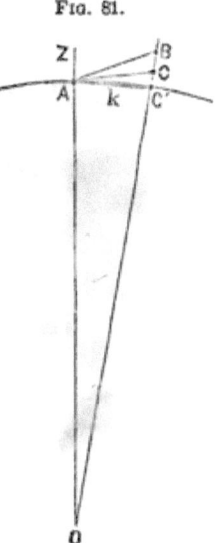

Fig. 81.

A correction for the curvature of the earth must therefore be made. It may be done in two ways: either by calculating C C′, and adding it to B C, obtained by formula [1]; or by calculating the angle C A C′, adding it to B A C, and then applying the formula [1] to the angle B A C′.

(101.) Correcting the Result. Expressing the distance by k, we have, by (14):

In feet $C C' = \dfrac{k^2}{2 R} = \dfrac{k^2}{2 \times 20888629} = 0.0000000023936 k^2$.

Then, calling A C B a right angle, we have:

$B C' = k \times \operatorname{tang.} B A C + 0.0000000023936 k^2$ in ft. [2.]

The arc A C′ and the straight lines A C′ and A C are all three approximately equal.

(102.) Correcting the Angle. The angle $C A C' = \tfrac{1}{2} A O C'$, the central angle, which is measured by the arc A C′ or k.

The length of the arc subtending one minute

$= \dfrac{2 \pi \times 20888629}{360 \times 60} = 6076$ ft.

Then, for any arc, k, the angle O in minutes

$$= \frac{k}{6076} = 0.0001646k;$$

and the angle C A C' (in minutes) $= 0.0000823k$.

Adding this to the observed angle, B A C, and calling A C' B a right angle, we have, by [1]:

$$B C' = k \text{ tang. } (B A C + 0.0000823k). \qquad [3.]$$

(103.) Correction for Refraction. The effect of refraction causes the angle actually observed to be, not C A B, but C A B', which will be designated by $a°$. For small distances, B and B' sensibly coincide. The correction for refraction may be made in two ways, as for curvature.

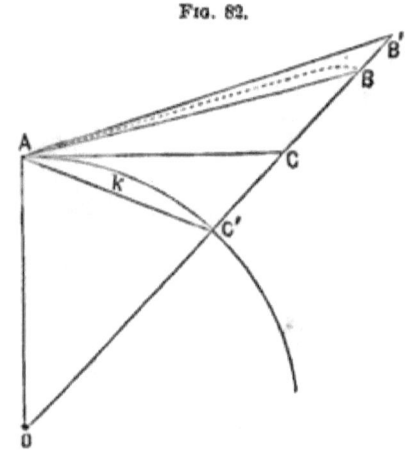

Fig. 82.

To correct the result by finding B B'. It varies very irregularly, with wind, barometer, temperature, etc.; but is usually taken, as an average, B B' $= 0.16$ C C'.

Subtracting this from the value of B C', in formula [2], it becomes B C' $= k$. tang. B' A C $+ 0.00000002k^2$. [4.]

To correct the observed angle. Subtract from it the angle B A B', which is about 0.16 of the angle C A C'.

This changes formula [3] to

$$B C' = k . \text{tang.} (B' A C + 0.000069k). \qquad [5.]$$

C.—*For Very Great Distances.*

(104.) Correction for Curvature. As before, there are two methods of making the correction.

For these distances we cannot consider the angle at C' a right angle. The triangle A B C gives

$$BC = k \cdot \frac{\sin. BAC}{\sin. B}.$$

To find the angle B, we have, in the triangle B A O,

$$B = 180° - (O + BAO),$$

$$B = 180° - (O + 90° + BAC),$$

$$B = 90° - (O + BAC);$$

Hence, sin. B = cos. (O + B A C).

Then, $BC = k \cdot \dfrac{\sin. BAC}{\cos. (O + BAC)}$,

and $BC' = BC + CC' = k \cdot \dfrac{\sin. BAC}{\cos. (O + BAC)} + 0.000000023936 k^2$

$$BC' = k \cdot \frac{\sin. BAC}{\cos. (BAC + 0.0001646k)} + 0.000000023936 k^2. \quad [6.]$$

Correcting the Angle. In the triangle A B C', getting expressions for the angles, and using the sine proportion, as before, in A B C, we have:

$$BC' = k \cdot \frac{\sin. (BAC + \frac{1}{2} O)}{\cos. (BAC + O)}.$$

$$BC' = k \cdot \frac{\sin. (BAC + 0.0000823k)}{\cos. (BAC + 0.0001646k)}. \quad [7.]$$

(105.) **Correction for Refraction.** Formula [6] becomes

$$BC' = k \cdot \frac{\sin. (BAC - 0.00001316k)}{\cos. (B'AC + 0.00015088k)} + 0.000000023936 k^2. \quad [8.]$$

SIMPLE ANGULAR LEVELLING. 67

Formula [7] becomes, diminishing B A C in both numerator and denominator by 0.08 of O,

$$B C' = k \cdot \frac{\sin. (B' A C + 0.00006918k)}{\cos. (B' A C + 0.0015148k)}. \qquad [9.]$$

(106.) **Reciprocal Observations for Cancelling Refraction.** Observe the reciprocal angles of elevation and depression from each point to the other. Call these angles a and β. Then:

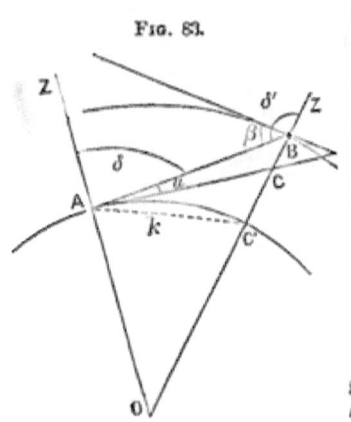

Fig. 83.

$$B C' = k \cdot \frac{\sin. \frac{1}{2} (a + \beta)}{\cos. \frac{1}{2} (a + \beta + O)}. \qquad [10.]$$

NOTE.—Angle O, in minutes $= 0.0001646k$.
Log. $0.0001646 = \overline{4}.2164298$.

When zenith distances are observed, they are denoted by δ and δ'. Then formula [10] becomes:

$$B C' = k \cdot \frac{\sin. \frac{1}{2} (\delta' - \delta)}{\cos. \frac{1}{2} (\delta' - \delta + O)} \qquad [10'.]$$

When O is very small, compared with the other angles, by neglecting it we have:

$$B C' = k \cdot \tang. \tfrac{1}{2} (a + \beta). \qquad [11.]$$

Using zenith distances, this becomes:

$$B C' = k \cdot \tang. \tfrac{1}{2} (\delta' - \delta). \qquad [11'.]$$

(107.) **Reduction to the Summits of the Signals.** Stations a and b cannot be seen from one another. Signals, Art. (240), are erected at each point, and from a the angle $B a C = A$ is observed; and from b the angle $A b D = B$. The heights of the signals above the instrument are h and h'.

Required the reduced angles a and β.

$$\left.\begin{array}{l} a = A - \dfrac{h \cdot \cos A}{k \cdot \sin 1''} \\[2ex] \beta = B - \dfrac{h' \cdot \cos B}{k \cdot \sin 1''} \end{array}\right\} \quad [12].$$

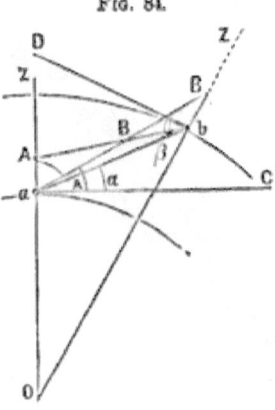

Fig. 84.

The difference is in seconds.

Usually, in such cases, zenith distances are taken, and the observed angles are called $\triangle$ and $\triangle'$. The reduced angles are δ and δ'. Then the formulas become:

$$\delta = \triangle + \frac{h \cdot \sin \triangle}{k \cdot \sin 1''}, \text{ and } \delta' = \triangle' + \frac{h' \cdot \sin \triangle'}{k \cdot \sin 1''}. \quad [13.]$$

The difference is in seconds.

Instead of h and h' some writers use dH and dH'; or dA and dA', meaning difference of height, and difference of altitude.

For great exactness, instead of using the mean radius of the earth to get O, the radius at the point of observation is used.

(108.) When the height of the signal above the instrument cannot be measured, if the signal be conical, like a spire, etc.,

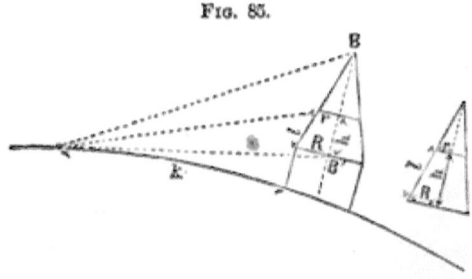

Fig. 85.

to find B B' we measure two diameters, $2R$ and $2r$, and the distance apart, h.

SIMPLE ANGULAR LEVELLING.

Then, $BB' = \dfrac{Rh}{R-r}$. [14.]

If the oblique distance l be measured instead of h, then

$$BB' = \dfrac{R}{R-r}\sqrt{[l+(R-r)][l-(R-r)]}.\quad [15.]$$

When the spire is very acute, then this method is inaccurate.

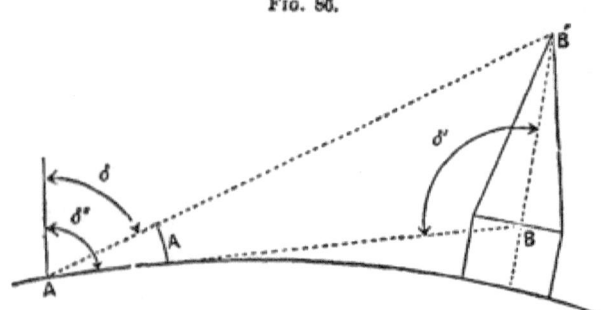

FIG. 86.

Take some point, A, and observe zenith distances, δ, δ'', and δ'. Then:

$$BB' = \dfrac{k \cdot \tan(\delta'' - \delta)}{\cos \tfrac{1}{2}(\delta' - \delta + O)} \quad [16.]$$

(109.) Levelling by the Horizon of the Sea. From an eminence, as B, sight to the sea horizon, and measure $\delta° =$ angle A B Z. Then:

FIG. 87.

$$BC' = \tfrac{1}{2} R \left(\dfrac{\sin. 1''}{1-n}\right)^2 (\delta° - 90°)^2 \left[1 + \tfrac{1}{4}\left(\dfrac{\sin. 1''}{1-n}\right)^2 (\delta° - 90°)^2\right]. \quad [17.]$$

$(\delta° - 90°)$ is to be reduced to seconds. It is equal to the angle of depression at B; n is the coefficient of refraction. It is taken at 0.08.

CHAPTER III.

COMPOUND ANGULAR LEVELLING.

The following problems may mostly be reduced to a combination of: first, determining the inaccessible distance to a point immediately under (or over) the point whose height is desired, and then using this distance to obtain that height.

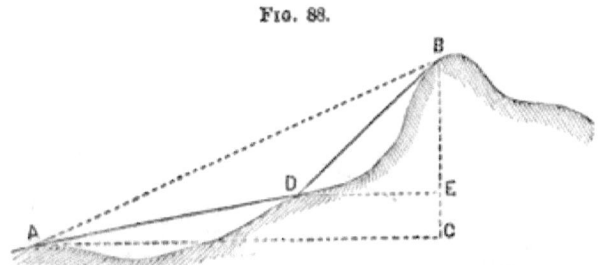

Fig. 88.

(110.) By Angular Co-ordinates in one Plane. Take two stations, A and D, in the same vertical plane with B. At A observe the angles of elevation of B and D. Measure A D. At D observe angle A D B. Then, in triangle A B D we get A B, and in triangle B A C we get B C.

$$BC = AD \cdot \frac{\sin. BDA \cdot \sin. BAC}{\sin. ABD}. \qquad [18.]$$

For great distances, the corrections for curvature and refraction are to be made as in last chapter.

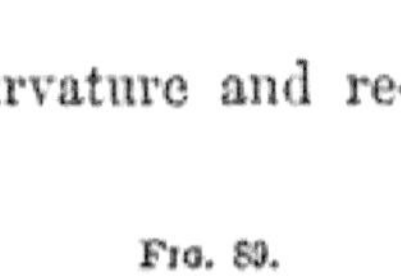

If A D be horizontal, the same formula applies; but there is one angle less to measure; since B A C = B A D. Formula [18] gives the height of B above A.

Fig. 89.

If the height of B above D, in Fig. 88, be desired, find B D in the triangle B A D, observe the angle of elevation of B from D, and then the desired height equals

$$BD \cdot \sin. BDE.$$

Otherwise, find height of D above A, and subtract it from B C.

(111.) By Angular Co-ordinates in several Planes. On irregular ground, when the distance between the two points is unknown, the operations for finding it by the various methods of L. S., Part VII., Chapter III., may be combined with the observation of vertical angles, thus:

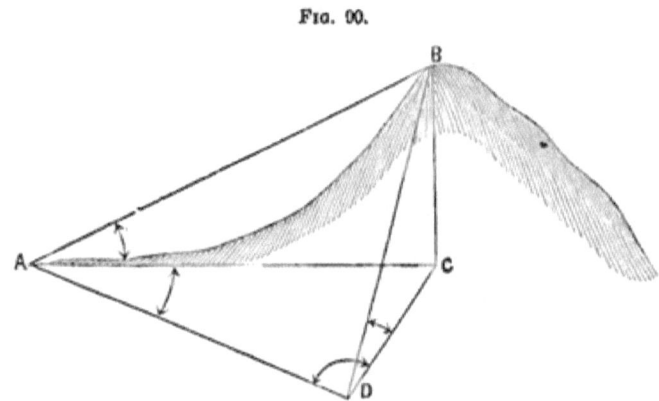

Fig. 90.

At A measure the vertical angle of elevation, B A C. Also measure the horizontal angle, C A D, to some point, D, and measure horizontally the distance, A D. At D measure the horizontal angle A D C. Then:

$$A C = A D \frac{\sin. A D C}{\sin. A C D}. \quad B C = A C . \tang. B A C =$$

$$B C = A D \frac{\sin. A D C . \tang. B A C}{\sin. A C D} \qquad [19.]$$

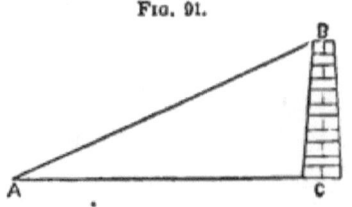

Fig. 91.

(112.) *Conversely.* The distance may be obtained when the height is known.

Let C B be a known height:

Then, $A C = C B . \tan. A B C.$

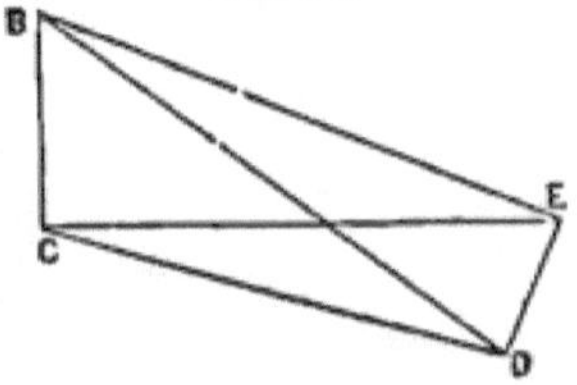

B C is a known height, and D E an inaccessible line in the same horizontal plane as C. Find C D and C E by the last method, and measure the horizontal angle E C D subtended at C by E D.

Then two sides and the included angle of a triangle are known, to find the third side.

Fig. 92.

PART III.

BAROMETRIC LEVELLING.

CHAPTER I.

PRINCIPLES AND FORMULAS.

(113.) **Principles.** The difference of the heights of two places may be determined by finding the difference of their depths below the top of the atmosphere in the same way as the comparative heights of ground under water are determined by the difference of the depths below the top of the water. The desired height of the atmosphere above any point, such as the top of a mountain, or the bottom of a valley, is determined by weighing it. This is done by trying how high a column of mercury or other liquid the column of air above it will balance; or what pressure it will exert against an elastic box containing a vacuum, etc., etc. Such instruments are called *Barometers*.

(114.) **Applications.** Since the column of mercury in the barometer is supported by the column of air above it, the mercury sinks when the barometer is carried higher, and *vice versa*.

The weight of any portion of air decreases from the surface of the earth to the assumed surface of the atmosphere. It has been found that, as the heights to which the barometer is carried increase in arithmetical progression, the weights of the column of air above the barometer, and consequently its readings, decrease in geometrical progression. Consequently, the difference of the heights of any two, not very distant, points on the earth's surface, is proportional to the difference of the logarithms of the readings of the barometer at those points; i. e., equal to this latter difference multiplied by some constant coefficient. This is found by experiment to be 60159, at the freezing point, or temperature of 32° F., the readings of the mercury being in inches, and the product, which is the difference of height, being in feet.

Several corrections are necessary.

(115.) Correction for Temperature of the Mercury. If the temperature of the mercury be different at the two stations, it is expanded at the one, and contracted at the other, to a height different from that which is due to the mere weight of the air above it.

Mercury expands about $\frac{1}{10000}$ of its bulk for each degree of F. Therefore, this fraction of the height of the column is to be added to the height of the colder column, or subtracted from the height of the warmer one, in order to reduce them to the same standard. A thermometer is therefore attached to the instrument in such a manner as to give the temperature of the mercury.

If a brass scale is used, the correction is $\frac{9}{100000}$ for each degree F.

(116.) Correction for Temperature of the Air. The warmer the air is, the lighter it is; so that a column of warm air of any height will weigh less than when it is colder. Consequently, the mercury in warm air falls less in ascending any height, and is higher at the place than it otherwise would be. Hence the height given by the preceding approximate result will be too small, and must be increased by $\frac{1}{491}$ part for each degree F. that the temperature of the air is above 32°. The effect of moisture in the air changes this fraction to $\frac{1}{450}$.

(117.) Other Corrections. For *very* great accuracy, we should allow for the variation of gravity, corresponding to the variation of latitude on either side of the mean. So, too, we should allow for the decrease of gravity corresponding to any increase of height of the place.

(118.) Rules for Calculating Heights by the Mercurial Barometer. 1. At each station read the barometer; note its temperature by the attached thermometer, and note the temperature of the air by a detached thermometer.

2. Multiply the height of the upper column by the difference of readings of the attached thermometer, and that by $\frac{9}{100000}$, and add the product to the upper column, if that be the colder, or subtract it, if that be the warmer. This gives the corrected height of the mercury.

3. Multiply the difference of the logarithms of the corrected heights of the mercury (i. e., the corrected upper one and the lower one) by 60159, and the product is the *approximate* difference of heights of the places in feet for the temperature of 32°.

4. Subtract 32° from the arithmetical mean of the temperatures of the detached thermometer; multiply the approximate altitude by this difference; divide the product by 450; add the quotient to the approximate altitude, and the sum is the corrected altitude.

(119.) Formulas. The rules just given are best expressed in formulas, thus:

	At Lower Station.	At Upper Station.
Height of Mercury.....................	H	h'
Temperature of Mercury..............	T	T'
Temperature of Air....................	t	t'

Calling the reduced height of mercury at upper station h, we have, by Rule 2:

$$h = h' + 0.00009 \, (T - T') \, h'. \qquad [1.]$$

(N. B. If T' is more than T, the product will be subtractive.)

Then, by Rule 3, we have:

Approx. height = 60159 (log. H − log. h).

By Rule 4, the correction for temperature of air

$$= \text{approx. height} \times \frac{t + t' - 64}{900}.$$

Adding this correction to the approximate height, and factoring the sum, we get:

$$\text{Corrected ht.} = 60159 (\log. H - \log. h)\left(1 + \frac{t + t' - 64}{900}\right) \quad [2].$$

(120.) **To Correct for Latitude.** Multiply the preceding result by 0.00265 . cos. 2L (L being the latitude), and add (algebraically) the product to the preceding result.

At 45°, correction is zero. At equator it is + 0.00265. At pole it is −0.00265.

To Correct for Elevation of the Place. Call the last corrected height x', and the height of the lower place above the level of the sea S, and add to x' this quantity:

$$\frac{x' + 52251}{20888629} + \frac{S}{10444315}.$$

(121.) **Final English Formula.** Combining the previous results into one formula, we get:

$$Ht. = 60159 (\log. H - \log. h) \begin{cases} \left(1 + \dfrac{t + t' - 64}{900}\right), \\ (1 + 0.00265 . \cos. 2L), \\ \left(1 + \dfrac{x' + 52251}{20888629} + \dfrac{S}{10444315}\right) \end{cases}$$
[3].

In this formula, the three quantities under each other are three factors.

Usually, only the first factor is required, and then we have formula [2]. Using the second also we correct for latitude; and using the third, for the elevation.

(122.) **French Formulas.** French barometers are graduated in French millimetres, each $= 0.03937$ inch., and the thermometer is centigrade, in which the freezing-point is zero, and boiling-point 100°:

$$a° \text{ Cent.} = (\tfrac{9}{5} a + 32)° \text{ F.}$$

Then, the French formula corresponding to [3] is the following (H and h' being in millimetres, and the temperatures centigrade):

$$h = h' \left(1 + \frac{T - T'}{6200}\right).$$

And the difference of heights in metres

$$= 18336 \,(\log. \text{H} - \log. h) \begin{cases} \left(1 + \dfrac{2(t + t')}{1000}\right), \\ (1 + 0.00265 \cdot \cos. 2\,\text{L}), \\ \left(1 + \dfrac{x' + 15926}{6366198} + \dfrac{S}{3183099}\right) \end{cases} \quad [4]$$

(123.) **Babinet's Simplified Formula, without Logarithms.**

h' is reduced to h, as before, viz.: $h = h'\left(1 + \dfrac{T - T'}{6200}\right).$

Then, the difference of heights in metres

$$= 16000 \cdot \frac{\text{H} - h}{\text{H} + h} \left(1 + \frac{2(t + t')}{1000}\right). \qquad [5.]$$

The heights are in millimetres and the temperatures centigrade.

Example. $H = 755.$ $h = 745$
$$t = 15° \quad t' = 10°$$

$$Ht. = 16000 \frac{10}{1500}\left(1 + \frac{50}{1000}\right) = 112 \text{ m.}$$

Correct result is 111.6 m.

This formula is a very near approximation for moderate heights.

Babinet's formula in English measures (the heights being in inches, and temperatures Fahrenheit) is in feet:

$$52494 \left(\frac{H - h}{H + h}\right)\left(1 + \frac{t + t' - 64}{900}\right). \qquad [6.]$$

(124.) **Tables.** These shorten the operations greatly. The most *portable* are in "Annuaire du Bureau des Longitudes." The most *complete* are Prof. Guyot's, published by the Smithsonian Institute at Washington.

(125.) **Approximations.** $\frac{1}{10}$ of an inch difference of readings at two places corresponds to about 90 feet difference of elevation. 1 millimetre difference of readings corresponds to about $10\frac{1}{2}$ metres difference of height, or about 34 feet.

This is correct near the freezing-point, and near the level of the sea. The height corresponding to any given difference of readings increases, however, with the temperature and with the height of the station. Thus, at 70° F., $\frac{1}{10}$ of an inch corresponds to an elevation of 95 feet; and 1 *mm.* at 30° Cent. corresponds to $11\frac{3}{4}$ metres, or about 40 feet.

CHAPTER II.

INSTRUMENTS.

(126.) Barometers made for levelling are called *Mountain Barometers*. They are either *cistern* barometers or *siphon* barometers. The best of the former is Fortin's, as improved by Prof. Guyot. (See Fig. 93.) This consists of a column of mercury contained in a glass tube, whose lower end is placed in a cistern of mercury. The tube is covered with a brass case, terminating at the top in a ring, A, for suspension, and at the bottom in a flange, B, to which the cistern is attached.

At C is a vernier by which the height of the mercury is read off. The zero of the scale is a small ivory point, P. The mercury in the cistern is raised or lowered, by means of the milled-headed screw O, till its surface is just in contact with the ivory point. At E is the attached thermometer which indicates the temperature of the mercury. When it is carried, the mercury is screwed up to prevent breaking the glass.[1]

In the siphon barometer, the cistern is dispensed with. The tube is turned up at the lower end, as shown in Fig. 94, and a small hole, at T, admits the air. The difference of heights of the mercury in the two branches of the tube is taken as the height of the mercurial column. It is enclosed in a brass case, and furnished with verniers, thermometers, etc., as in the preceding form. It is carried inverted, to avoid breaking.

The best is Gay-Lussac's, improved by Bunten.

---

[1] For a complete description, see Tenth Annual Report of Smithsonian Institute.

(127.) The Aneroid Barometer. This is a thin box of corrugated copper, exhausted of air. When the air grows heavier, the box is compressed; and when the air grows lighter it is

Fig. 95

expanded by a spring inside. This motion is communicated by suitable levers to the index-hand, on the face, which indicates the pressure of the atmosphere, the face being graduated to correspond with a common barometer.

It is much used on account of its portability, but is not as reliable as the mercurial barometer.

(128.) The temperature at which water boils varies with the pressure of the atmosphere, and therefore decreases in ascending heights. Then a thermometer becomes a substitute for a barometer.

Approximately, each de-

Temperature of Boiling Water.	Corresponding Barometer Readings.
213°	30".522
212°	29".922
211°	29".331
210°	28".751
209°	28".180
208°	27".618

gree of difference (Fahr.) corresponds to about 550 feet difference of elevation, subject to the usual barometric corrections for the temperature of the air. For minute tables, see Guyot's.

(129.) Accuracy of Barometric Observations. This increases with the number of repetitions of them, the mean being taken. With great skill and experience they may be depended upon to a very few feet.

PROFESSOR GUYOT'S RESULTS.

Heights found by the Barometer.	Corresponding Heights found by the Spirit-Level.
6707 feet.	6711 feet.
2752 "	2752 "
6291 "	{ 6285 " { 6293 "

(130.) The observations at the two places, whose difference of heights is to be determined, should be taken simultaneously at a series of intervals previously agreed upon, the distance apart of the places being as short as possible. Distant places should be connected by a series of intermediate ones.

PART IV.

TOPOGRAPHY.

INTRODUCTION.

(131.) Definition. Topography is the complete determination and representation of any portion of the surface of the earth, embracing the relative position and heights of its inequalities; its hills and hollows, its ridges and valleys, level plains, slopes, etc., telling precisely where any point is, and how high it is.

It therefore determines the three coördinates of any point; the horizontal ones by surveying, and the vertical one by levelling.

The results of these determinations are represented in a conventional manner, which is called "topographical mapping."

The difficulty is, that we see hills and hollows in *elevation*, while we have to represent them in *plan*.

(132.) Systems. Hills are represented by various systems:
1. By level contour-lines, or horizontal sections.
2. By lines of greatest slope, perpendicular to the former.
3. By shades from vertical light.
4. By shades from oblique light.

The most usual method is a combination of the first, second, and third systems.

CHAPTER I.

FIRST SYSTEM.

BY HORIZONTAL CONTOUR-LINES.

(133.) General Ideas. Imagine a hill to be sliced off by a number of equidistant horizontal planes, and their intersections with it to be drawn as they would be seen from above, or horizontally projected on the map, as in Fig. 96. These are "contour-lines."

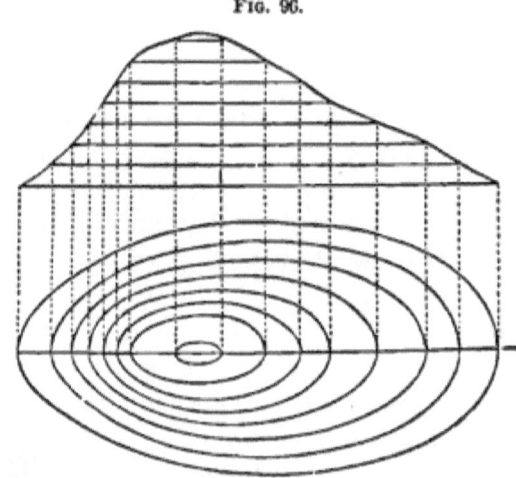

Fig. 96.

They are the same lines as would be formed by water surrounding the hill, and rising one foot at a time (or any other height), till it reached the top of the hill. The edge of the

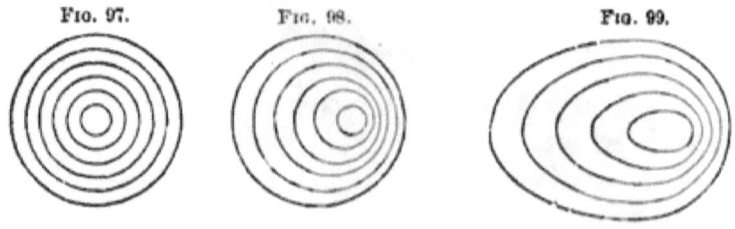

Fig. 97.   Fig. 98.   Fig. 99.

water, or its shore, at each successive rise, would be one of these horizontal contour-lines. It is plain that their nearness or distance on the map would indicate the steepness or gentleness of the slopes. A right cone would thus be represented by a series of concentric circles, as in Fig. 97; an oblique cone, by circles not concentric, but nearer to each other on the steep side than on the other, as in Fig. 98; and by a half-egg, somewhat as in Fig. 99.

(134.) **Plane of Reference.** The horizontal plane on which the contour-lines are projected, and to which they are referred, is called the "plane of reference." This plane may be assumed in any position, and the distance of the contour-lines above or below it is noted on them. It is usually best to assume the position of the plane of reference lower than any point to be represented; so that all the contour-lines will be above it, and none of them have minus signs. (See Art. 59.)

(135.) **Vertical Distances of the Horizontal Sections.** These depend on the object of the survey, the population of the country, the irregularity of the surface, and the scale of the map. In mountainous districts they may be 100 feet apart. On the United States Coast Survey they are 20 feet. For engineering purposes, 5 feet, or less. One rule is to make the distance in feet equal to the denominator of the ratio of the scale of the map, divided by 600.

(136.) **Methods for determining Contour-Lines.** They are of two classes: 1. Determining them on the ground at once; 2. Determining the highest and lowest points, and thence deducing the contour-lines.

First Method.

(137.) **General Method.** Determine one point at the desired height of one line, as in Art. (81); and then "locate" a line at that level, as in Art. (84).

The "reflected hand-level," or "reflecting-level," or "water-level," are sufficiently accurate between "bench-marks" not very distant.

One such line having been determined, a point in the next higher or the next lower one is fixed, and the preceding operations repeated.

(138.) On a long, narrow Strip of Ground, such as that required for locating a road: Run a section across it at every ¼ or ½ mile, about in the line of greatest slope. Set stakes on these sections at the heights of the desired contour-lines, and then get intermediate points at these heights between the stakes. These sections *check* the levels.

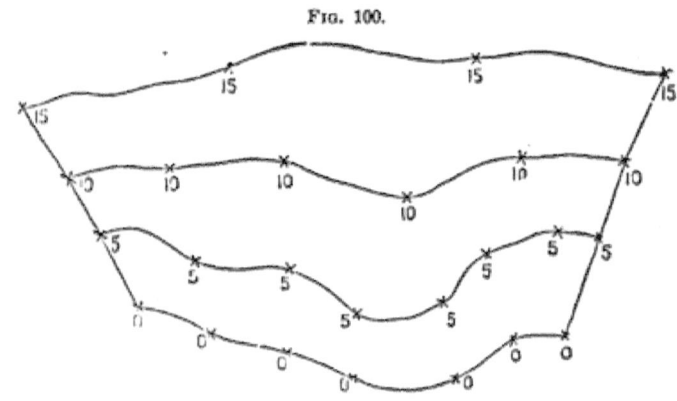

Fig. 100.

(139.) On a Broad Surface. Level around it setting-stakes, at points of the desired height, and then run sections across it, and from them obtain the contour-lines as before.

The external lines here serve as checks to the cross-lines.

(140.) Surveying the Contour-Lines. The contour-lines thus found may be surveyed by any method. If they are long, and not very much curved, the compass and chain and the method of "progression" is best. (See L. S. 220.) If they are curved irregularly, the method of radiation is best. When straight lines exist among them, such as fences, etc., or can conveniently be established, then rectangular coördinates are most convenient.

(**141.**) **Contouring with the Plane-Table.**[1] It is used to map the points as soon as obtained, thus: Range out an approximately level line, and on it set equidistant stakes. At these stakes range out perpendiculars to the line, and set up several stakes on them for the alignment of the rodman. Draw these lines on the plane-table. Set up and "orient" (L. S. 456) the table on the ground. Send the rod along one of the perpendiculars till it comes to a point of the right height. Then sight to it with the alidade, and its edge will cut the corresponding line on the table at the correct place on the plat. So for the other perpendiculars.

Second Method.

(**142.**) **General Nature.** This method consists in determining the heights and positions of the principal points, where the surface of the ground changes its slope in degree or in direction, i. e., determining all the highest and lowest points and lines, the tops of the hills and bottoms of the hollows, ridges and valleys, etc., and then, by proportion or interpolation, obtaining the places of the points which are at the same desired level. The heights of the principal points are found by common levelling, and their places fixed as in Art. (151).

The first method is more *accurate*. The second is more *rapid*.

(**143.**) **Irregular Ground.** When the ground has no very marked features, run lines across it in various directions, and level along them, taking heights at each change of slope, just as in taking sections for profiles.

Otherwise, thus: Set stakes on four sides of the field, so as to enclose it in a rectangle, if possible, as in Fig. 101. Place the stakes equidistant, so that the imaginary visual lines connecting them would divide the surface into rectangles. Send the rod along one of these lines till it gets in the range of a cross-one, and observe to it there. Put down the observed heights of these points at the corresponding points on

---

[1] For description and method of using the Plane-table, see L. S. Part VIII.

the plat, on which these lines have been drawn. The contour-lines are determined as in Art. (146).

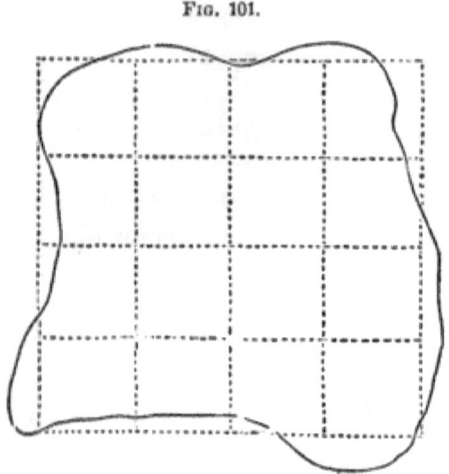

Fig. 101.

(144.) **On a Single Hill.** Proceed thus: From its top, range lines down the hill, in various directions, and take their bearings. Set stakes on them at each change of slope, and note the heights and distances of these stakes from the starting-point, and plat their places. The contour-lines are then put in as in Art. (146).

With a transit, the heights of the points could be determined by vertical angles; and also their distances with stadia-hairs, their directions being given by the horizontal circle of the transit. The French use for this purpose a "levelling compass."

(145.) **For an Extensive Topographical Survey.** Proceed thus: Set up, and get the height of the cross-hairs from some bench-mark, and get the heights of high and low prominent points all around. Then go beyond these points and set up again. Sight to one of these known points as a "turning-point," and get the heights of all the points now in sight, as before. Then go beyond these again, and so on. The places of these new points are fixed as before.

(**146.**) **Interpolation.** The heights and the places of the principal points being determined, by either of the pre-

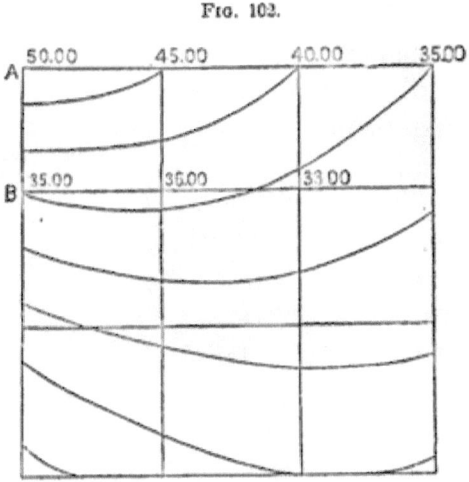

Fig. 102.

ceding methods, points of any intermediate height, corresponding to any desired contour-curve, are obtained by proportion.

If, in Fig. 102, the heights of the intersection of the lines being found, as in Art. (143), and their distance apart being 100 feet, it is required to construct contour-curves whose difference of heights is 5 feet: Taking for example the one whose height is 45 feet, we see it must fall between the points A and B, whose heights are 50 feet and 35 feet; and its distance from A will be found by the proportion, as 15 is to 5 so is 100 to the required distance. So on for any number of points. To save the labor of continually calculating the fourth proportional, a scale of proportion may be constructed.

(**147.**) **Interpolating with the Sector.** (L. S. 52.) This is the easiest way. The problem is: having given on a plat two points of known height, to interpolate between them a point of any desired intermediate height.

Take in the dividers the distance between the given points on the plat; open the sector so that this distance shall just reach between numbers, on the scale marked L, corresponding to the difference of the heights of the two given points; i. e., from 6 to 6, or 7 to 7, and so on. The sector is then *set* for all the interpolations between these two points.

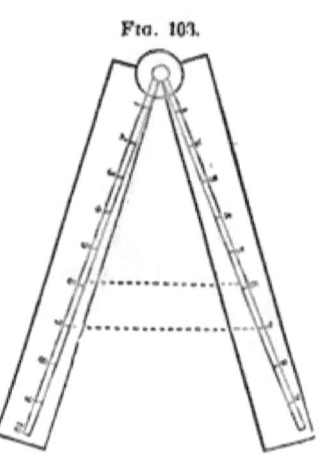

Fig. 103.

Then note the difference of height between the desired point and one of the given points, and extend the dividers between the corresponding numbers on the scale. This opening will be the distance to be set off on the plat from the given point to the desired point.

(148.) **Ridges and Thalwegs.** The general character of the surface of a country is given by two sets of lines: the *ridge-lines*, or *water-shed lines;* and *the "thalwegs,"* or *lowest lines of valleys.*

The former are lines which divide the water falling upon them, and from which it passes off on contrary sides. They are the lines of least slope when looking along them from above downward; and they are the lines of greatest slope when looking from below upward. They can therefore be readily determined by the slope-level, etc. They are the lines of *least* zenith distances when viewed from either direction.

On these lines are found all the projecting or protruding bends of the contour-lines, convex toward the lower ground, as shown in Fig. 104.

The second set of lines, or the "thalwegs," are the converse of the former. They are indicated by the water-courses which follow them or occupy them. They are the lines of greatest slope when looked at from above, and of least slope when looked at from below. They are the lines of *greatest* zenith distance when viewed from either direction.

On these lines are the receding or reëntering points of the contour-curves, concave toward the lower ground.

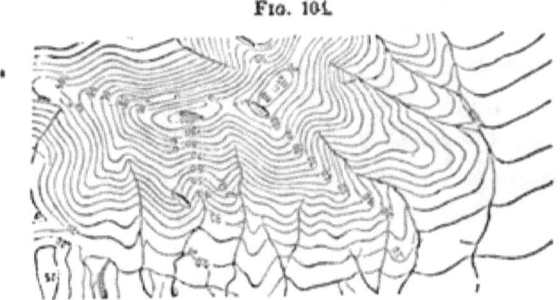

Fig. 104.

The general system of the surface of a country is most easily characterized by putting down these two sets of lines, and marking the changes of slope, especially the beginning and the end.

The most important points to be determined are:
1. At the top and bottom of slopes.
2. At the changes of slopes in degree.
3. On the water-shed lines, and on the thalwegs.
4. On "cols," or culminating points of passes.

(149.) **Forms of Ground.** It will be found on the inspection of a "contour-map" (which shows ground much more plainly to the eye than does the ground itself), that its infinite variety of form may, for the purposes of the engineer, be reduced to five: 1. Sloping down on all sides; i. e., a hill, Fig. 105.

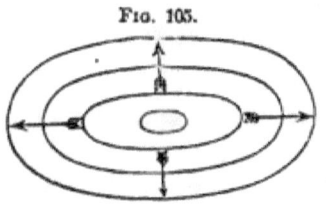

Fig. 105.

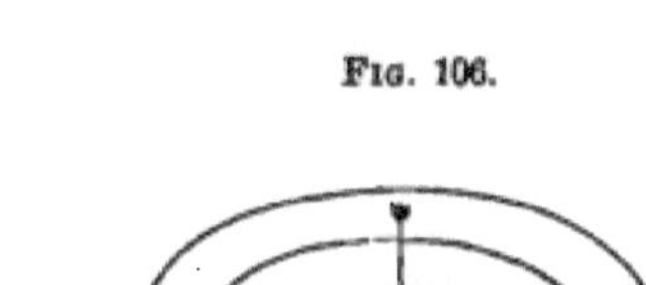

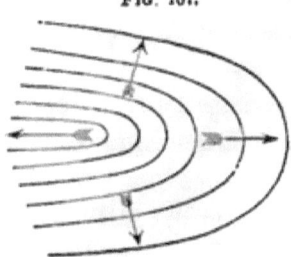

Fig. 106.

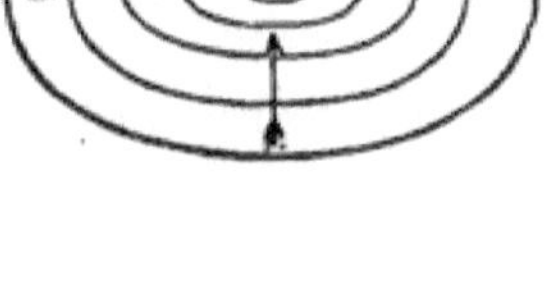

Fig. 107.

FIRST SYSTEM. 91

2. Sloping up on all sides; i. e., a hollow, Fig. 106.

3. Sloping down on three sides and up on one; i. e., a *croupe*, or shoulder, or promontory, the end of a ridge or water-shed line, Fig. 107.

4. Sloping up on three sides and down on one; i. e., a valley, or "thalweg," Fig. 108.

5. Sloping up on two sides and sloping down on two, alternately; i. e., a "pass," or "*col*," or "saddle," Fig. 109.

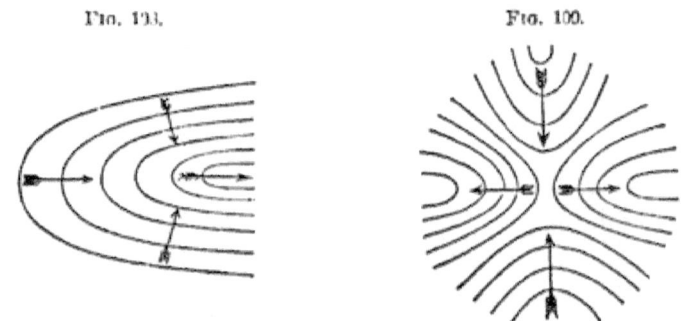

Fig. 108.   Fig. 109.

[NOTE.—The arrows in the figures indicate the direction in which water would run.]

(150.) Sketching Ground by Contours. A valuable guide is, the observation that the lines are perpendicular to the water-shed lines and thalwegs. Note especially the contour-lines at the bottoms of hills and ridges, and at the tops of hollows and valleys, putting them down, in their true relative positions and distances, to an estimated scale.

On a long slope or hill, draw first the bottom contour-line, and the top one; and then the middle one; and afterward interpolate others. Remember that two of them can never meet, except on a perpendicular face; and that, if one of them passes entirely around a hill or hollow, it will come back to its starting-point. Hold the field-book so that the lines on it have their true direction.

(151.) Ambiguity. In contour-maps of ground, if the heights of the contour-lines are not written upon them, it may be doubtful which are the highest and lowest; which are ridges and which valleys, etc.

1. Numbers remove this.

2. The water-courses show the slopes. If there are none, put some in, in the thalwegs of a rough sketch.

3. Put hatchings on the lower sides of the contour-lines, as if water were draining off.

4. Tint the valleys and low places.

(152.) Conventionalities. Sometimes the spaces between contour-lines are colored with tints of Indian-ink, sepia, etc., increasing in darkness as the depth increases.

Ground under water is commonly so represented, beginning at the low-water line and covering the space to the six-feet-deep contour-line with a dark shade of Indian-ink; then a lighter shade from 6 to 12; a still lighter from 12 to 18, and the lightest from 18 to 24.

Greater depths are noted in fathoms and fractions.

(153.) Applications of Contour-Lines. They have many important uses besides their representation of ground:

1. To obtain vertical sections; i. e., profiles.

2. To obtain oblique sections.

3. To locate roads.

4. To calculate excavation and embankment. Consider the contour-lines to represent sections of the mass by horizontal planes. Then each slice between them will have its contents equal, approximately, to half the sum of its upper and lower surfaces multiplied by the vertical distance apart of the sections. The areas may be obtained as in L. S. (74) and (124). This method is used to get the cubic contents of a hill to be cut away; of a hollow to be filled up; of a great reservoir in a valley, either only projected, or full of water, etc.

(154.) Sections by Oblique Planes. This method was much used by the old military topographers. It is picturesque, but not precise. The cutting-planes are parallel, and may make any angle with the horizon.

CHAPTER II.

SECOND SYSTEM.

BY LINES OF GREATEST SLOPE.

(155.) Their Direction. It is that which water would take in running down a slope. They are drawn perpendicularly to the contour-lines, and are the "lines of greatest slope." They are called "hatchings."

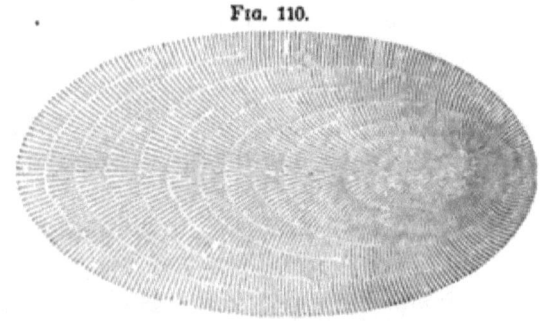

Fig. 110.

Fig. 110 represents an oval hill by this system.

(156.) Sketching Ground by this System. This is rapid and effective, but not precise. In doing this, hold the book to correspond with your position on the ground, and always draw toward you. If at the top of a hill, begin by drawing lines from the bottom, and *vice versa*. The hatchings are guided by contour-lines lightly sketched in.

(157.) Details of Hatchings. They must be drawn very truly perpendicular to the contour-lines. But if the contour-lines are not parallel, the hatchings must curve. When the contours are very far apart, as on nearly level ground, then pencil in intermediate ones.

Hatchings in adjoining rows should not be continuous, but "break joints," to indicate the places of the contour-lines, which are usually pencilled in to guide the hatchings, and then rubbed out. The rows of hatchings must neither overlap nor separate, and the lines should be made slightly tremulous. When they are put in without contour-lines to guide them, take care never to let two rows run into one; for the breaks between the rows represent contour-lines, and two contour-lines of different heights can never meet except on a vertical surface.

CHAPTER III.

THIRD SYSTEM.

BY SHADES FROM VERTICAL LIGHT.

(158.) Degree of Shade. The steeper the slope is, the less light it receives, in the inverse ratio of its length; i. e., inversely as the secant of the angle a which it makes with the horizon, or directly as $\cos a$. Then the ratio of the black to the white is, :: $1 - \cos a : \cos a$.

Fig. 111.

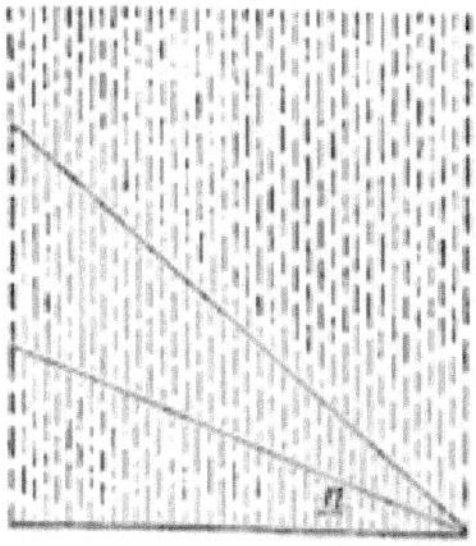

In practice, the difference of shade is much exaggerated.

Tables have been prepared by various nations, establishing the ratio of black and white.

The proper degree of shade may be given to the hills and hollows on the map by various means.

(159.) Shades by Tints. Indian-ink, or sepia, is used. The shades are put on with proper darkness, according to a previously-prepared "diapason of tints." The tints are made light for gentle slopes, and dark for steep slopes, in a constant ratio, a slope of 60° being quite black, one of 30° a tint midway between that and white, and so on. The edges at the top and bottom are softened off with a clean brush. This is rapid and effective, but not very definite or precise, except in combination with contour-lines.

(160.) **Shades by Contour-Lines.** This is done by making the contour-lines more numerous; i. e., interpolating new ones between those first determined. One objection to this is confusion of these lines with roads.

(161.) **Shades by Lines of Greatest Slope.** The lines of steepest slope, i. e., the hatchings between the contours, have their thickness and distance apart made proportional to the steepness of the slope, in some definite ratio. This is the most usual method.

The tints may be produced by varying the thickness of the hatchings, or their distance apart. Both are usually combined.

(162.) **The French Method.** In this the degree of inclination is indicated by varying the distances between the centres of the hatchings. The rule is: *the distance between the centres of the lines shall equal $\frac{2}{100}$ of an inch, plus $\frac{1}{4}$ of the denominator of the fraction denoting the declivity* (i. e., tangent of the angle made by the surface of the ground with the plane of reference) *expressed in hundredths of an inch.*

The lines are made heavier as the slope is steeper, being fine for the most gentle slopes, and increasing in breadth till the blank space between them equals $\frac{1}{2}$ the breadth of the lines.

Only slopes of from $\frac{1}{1}$ to $\frac{1}{64}$ inclusive are represented by this method.

(163.) **The German, or Lehmann's Method.** He uses nine grades for slopes from 0° to 45°, the first being white and the last black. For the intermediate slopes, he makes the white to the black in the following proportion:

The white : the black :: 45° − angle of slope : angle of slope.

For example, for 30°:

light : shade :: 45° − 30° : 30° :: 1 : 2.

Hence, the space between the strokes is to their thickness, as 45° minus the angle of the slope is to the angle of the slope.

96 LEVELLING, TOPOGRAPHY, AND HIGHER SURVEYING.

Slopes steeper than 45° are represented by short, heavy lines,

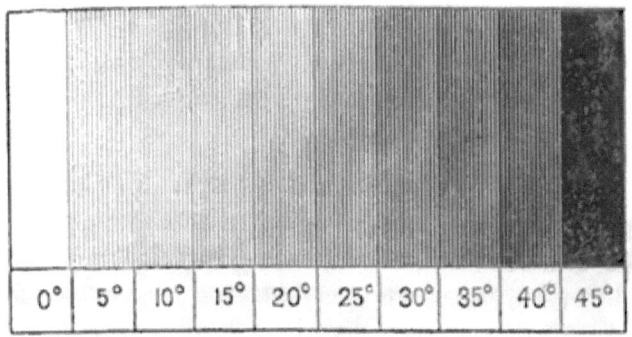

FIG. 112.

parallel to the contour-lines, as shown in the upper right-hand corner of Fig. 113—a hill drawn by Lehmann's Method.

FIG. 113.

(164.) Another Diapason of Tints:

Slope.	2½°	5°	10°	15°	25°	35°	45°	60°	75°
Black.	1	2	3	4	5	6	7	8	9
White.	10	9	8	7	6	5	4	3	2

This distinguishes gentle slopes better. It makes them darker, and the steeper slopes lighter, and provides for slopes beyond 45°. To use this standard, make it on the edge of a strip of paper, and apply that to the map in various parts, and draw a few lines corresponding to the slope of those parts; then fill up the intervening portions with suitable gradations. The angle of the slope is known from the map, since its tangent equals the vertical distance between the contours, divided by the horizontal distance. A scale can be made for any given vertical distance.

FOURTH SYSTEM.

BY SHADES PRODUCED BY OBLIQUE LIGHT.

(165.) Light is supposed to fall from the upper left-hand corner, as in drawing an "elevation," although the map is in plan. Then slopes facing the light will have a light tint, and those on the opposite side a dark tint.

This is picturesque, but not precise. It gives apparent "relief" to the ground, but does not show the degree of steepness.

The shades may be produced, as in the last method, by any means—tints, contours, or hatchings.

By making a map with contour-lines, and shaded obliquely, it will be both effective and precise.

CHAPTER IV.

CONVENTIONAL SIGNS.

(166.) **Signs for Natural Surface.** *Sand* is represented by fine dots made with the point of the pen; *gravel*, by coarser dots. *Rocks* are drawn in their proper places, in irregular angular forms, imitating their true appearance as seen from above. The nature of the rocks, or the *geology* of the country, may be shown by applying the proper colors, as agreed on by geologists, to the back of the map, so that they may be seen by holding it up against the light, while they will thus not confuse the usual details.

(167.) **Signs for Vegetation.** *Woods* are represented by scolloped circles, irregularly disposed, imitating trees seen "in plan," and closer or farther apart according to the thickness of the forest. It is usual to shade their lower and right-hand sides, and to represent their shadows, as in the figure, though, in strictness, this is inconsistent with the hypothesis of vertical light, usually adopted for "hill-drawing." For pine and similar forests, the signs may have a star-like form, as on the right-hand side of the figure. Trees are sometimes drawn "in elevation," or sideways, as usually seen. This makes them more easily recognized, but is in utter violation of the principles of mapping in horizontal projection, though it may be defended as a pure convention. *Orchards* are represented by trees arranged in rows. *Bushes* may be drawn like trees, but smaller.

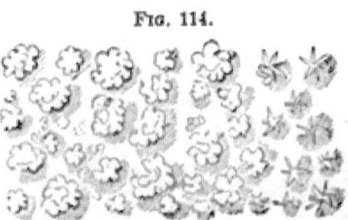

Fig. 114.

Grass-land is drawn with irregularly scattered groups of short lines, as in the figure, the lines being arranged in odd numbers, and so that the top of each group is convex, and its bottom horizontal or parallel to the base of the drawing. *Meadows* are sometimes represented by pairs of diverging lines (as on the right of the figure),

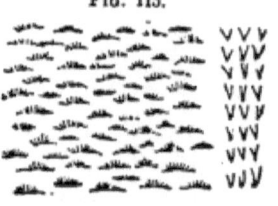

Fig. 115.

which may be regarded as tall blades of grass *Uncultivated* land is indicated by appropriately intermingling the signs for grass-land, bushes, sand, and rocks. *Cultivated* land is shown by parallel rows of broken and dotted lines, as in the figure, representing furrows. *Crops* are so temporary that signs for them are unnecessary, though often used. They are usually imitative, as for cotton, sugar, tobacco, rice, vines, hops, etc. *Gardens* are drawn with circular and other beds and walks.

Fig. 116.

(168.) **Signs for Water.** The *Sea-coast* is represented by drawing a line parallel to the shore, following all its windings

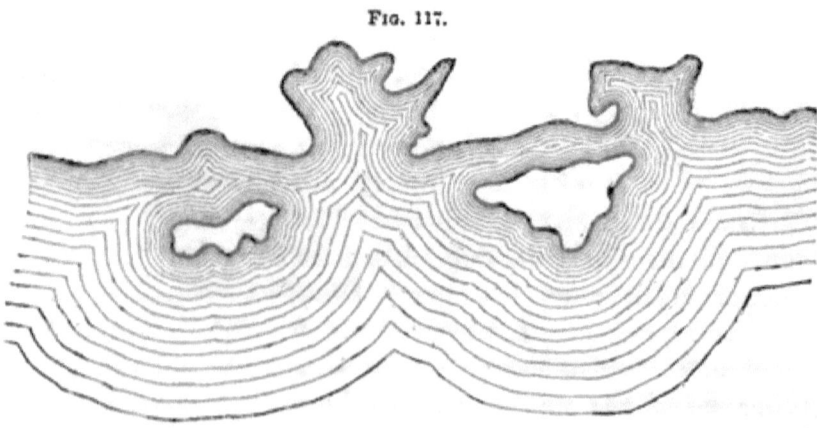

Fig. 117.

and indentations, and as close to it as possible; then another parallel line a little more distant; then a third still more distant, and so on, as in Fig. 117. If these lines are drawn from the low-tide mark, a similar set may be drawn between that and the high-tide mark, and dots, for sand, be made over the included space.

Rivers have each shore treated like the sea-shore, as in Fig. 118.

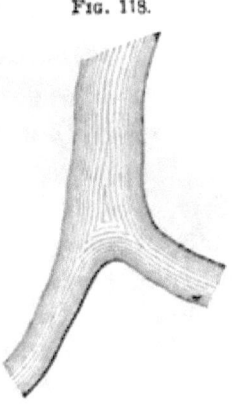

Fig. 118.

Brooks would be shown by only two lines, or one, according to their magnitude.

Fig. 119.

Ponds may be drawn like sea-shores, or represented by parallel horizontal lines ruled across them. *Marshes* and *Swamps* are represented by an irregular intermingling of the preceding sign with that for grass and bushes, as shown in Fig. 119.

(169.) **Colored Topography.** The conventional signs which have been described, as made with the pen, require much time and labor. Colors are generally used by the French as substitutes for them, and combine the advantages of great rapidity and effectiveness. Only three colors (besides Indian-ink) are required, viz.: Gamboge (yellow), Indigo (blue), and Lake (scarlet), Sepia, Burnt Sienna, Yellow Ochre, Red Lead, and Vermilion, are also sometimes used. The last three are difficult to work with. To use these paints, moisten the end of a cake and rub it up with a drop of water, afterward diluting this to the proper tint, which should always be light and delicate. To cover any surface with a uniform flat tint, use a large camel's-hair or sable brush, keep it always moderately full, incline the board toward you, previously moisten the paper with clean water if the outline is very irregular, begin at the top of the surface, apply a tint across the upper part, and continue it downward, *never letting the edge dry.* This last is the secret of a smooth tint. It requires rapidity in returning to the beginning of a tint to continue it, and dexterity in following the outline. *Marbling*, or variegation, is produced by having a brush at each end of a stick, one for each color, and applying first one, and then the other, beside it before it dries, so that they may blend, but not mix, and produce an irregularly-clouded appearance. Scratched parts of the paper may be painted over by first applying strong alum-water to the place.

The conventions for Colored Topography, adopted by the French military engineers, are as follows: WOODS, *yellow;* using gamboge and a very little indigo. GRASS-LAND, *green;* made of gamboge and indigo. CULTIVATED LAND, *brown;* lake, gamboge, and a little Indian-ink; "burnt sienna" will answer. Adjoining fields should be slightly varied in tint. Sometimes furrows are indicated by strips of various colors. GARDENS are represented by small rectangular patches of brighter *green* and *brown.* UNCULTIVATED LAND, marbled *green* and light *brown.* BRUSH, BRAMBLES, etc., marbled *green* and *yellow.* HEATH, FURZE, etc., marbled *green* and *pink.* VINEYARDS, *purple;* lake and indigo. SANDS, a light *brown;* gamboge and lake; "yellow ochre" will do. LAKES and RIVERS, light *blue*, with a darker tint on their upper and left-hand sides. SEAS, dark *blue*, with a little yellow added. MARSHES, the *blue* of water, with spots of grass, *green*, the touches all lying horizontally. ROADS, *brown;* between the tints for sand and cultivated ground, with more Indian-ink. HILLS, *greenish brown;* gamboge, indigo, lake, and Indian-ink. WOODS may be finished up by drawing the trees as in Art. (167), and coloring them green, with touches of gamboge toward the light (the upper and left-hand side), and of indigo on the opposite side.

(170.) **Signs for Miscellaneous Objects.** Too great a number of these will cause confusion. A few leading ones will be given:

102   LEVELLING, TOPOGRAPHY, AND HIGHER SURVEYING.

Signal of survey,	△	120	Saw mill,		129
Telegraph,		121	Wind mill,		130
Court house,		122	Steam mill,		131
Post office,		123	Furnace,		132
Tavern,		124	Woollen factory,		133
Blacksmith's shop,		125	Cotton factory,		134
Guide board,	†	126	Glass works,		135
Quarry,		127	Church,		136
Grist mill,		128	Grave yard,		137

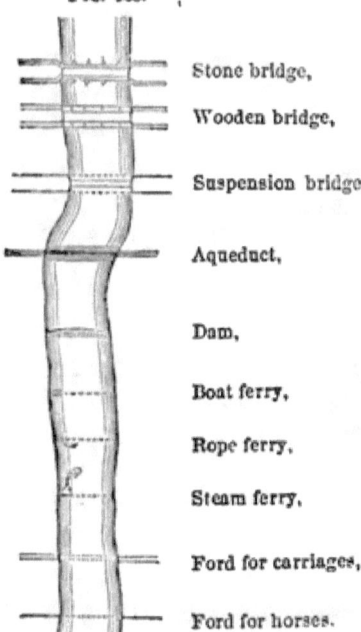

FIG. 138.

Stone bridge,
Wooden bridge,
Suspension bridge,
Aqueduct,
Dam,
Boat ferry,
Rope ferry,
Steam ferry,
Ford for carriages,
Ford for horses.

An ordinary house is drawn in its true position and size, and the ridge of its roof shown, if the scale of the map is large enough. On a very small scale, a small shaded rectangle represents it. If colors are used, buildings of masonry are tinted a deep crimson (with lake), and those of wood with Indian-ink. Their lower and right-hand sides are drawn with heavier lines. Fences of stone or wood, and hedges, may be drawn in imitation of the realities; and, if desired, colored appropriately.

Mines may be represented by the signs of the planets, which were anciently associated with the various metals. The signs here given represent respectively:

| Gold. | Silver. | Iron. | Copper. | Tin. | Lead. | Quicksilver. |

A large black circle, ●, may be used for Coal.

Boundary-lines, of private properties, of townships, of counties, and of States, may be indicated by lines formed of various combinations of short lines, dots, and crosses, as below.[1]

— — — — — — — — — — — — — — —

. . . .

—.—.—.—.—.—.—.—.—.—.—.—.—

+ + + + + + + + + + + + + + + + +

(171.) Scales. The scale to which a topographical map should be drawn, depends on several considerations. The principal ones are these: It should be large enough to express all necessary details, and yet not so large as to be unwieldy. The scale should be so chosen that the dimensions measured on the ground can be easily converted, without calculation, into the corresponding dimensions on the map.

In the United States Engineer service, the following scales are prescribed:

General plans of buildings, 1 inch to 10 feet (1 : 120).
Maps of ground, with horizontal curves one foot apart, 1 inch to 50 feet (1 : 600).
Topographical maps, one mile and a half square, 2 feet to one mile (1 : 2,640).
Do. comprising three miles square, 1 foot to one mile (1 : 5,280).
Do. between four and eight miles square, 6 inches to one mile (1 : 10,560).
Do. comprising nine miles square, 4 inches to one mile (1 : 15,840).
Maps not exceeding 24 miles square, 2 inches to one mile (1 : 31,680).
Maps comprising 50 miles square, 1 inch to one mile (1 : 63,360).
Maps comprising 100 miles square, ½ inch to one mile (1 : 126,720).
Surveys of roads, canals, etc., 1 inch to 50 feet (1 : 600).

On the admirable United States Coast Survey, all the scales are expressed fractionally and decimally. The surveys are generally platted originally on a scale of one to ten or twenty thousand, but in some instances the scale is larger or smaller.

---

[1] Very minute directions for the execution of the details of topographical mapping, are given in Lieutenant R. S. Smith's "Topographical Drawing." Wiley, New York.

These original surveys are reduced for engraving and publication, and, when issued, are embraced in three general classes: 1°, Small Harbor Charts; 2°, Charts of Bays, Sounds; and 3°, of the Coast General Charts.

The scales of the first class vary from 1 : 10,000 to 1 : 60,000, according to the nature of the harbor and the different objects to be represented.

Where there are many shoals, rocks, or other objects, as in Nantucket Harbor and Hell-Gate, or where the importance of the harbor makes it necessary, a larger scale of 1 : 5,000, 1 : 10,000, and 1 : 20,000, is used. But where, from the size of the harbor, or its ease of access, a smaller one will point out every danger with sufficient exactness, the scales of 1 : 40,000 and 1 : 60,000 are used, as in the case of New-Bedford Harbor, Cat, and Ship Island Harbor, New Haven, etc.

The scale of the second class, in consequence of the large areas to be represented, is usually fixed at 1 : 80,000, as in the case of New-York Bay, Delaware Bay and River. Preliminary charts, however, are issued, of various scales from 1 : 80,000 to 1 : 200,000.

Of the third class, the scale is fixed at 1 : 400,000 for the general chart of the coast from Gay Head to Cape Henlopen, although considerations of the proximity and importance of points on the coast may change the scales of charts of other portions of our extended coast.

PART V.

UNDERGROUND OR MINING SURVEYING.

(172.) It has three objects:

1. To determine the directions and extent of the present workings of a mine.

2. To find a point on the surface of the ground from which to sink a shaft, to meet a desired spot of the underground workings.

3. To direct the underground workings to meet a shaft or any other desired point.

It attains these objects by a combination of surveying and levelling.

CHAPTER I.

SURVEYING AND LEVELLING OLD LINES.

(173.) **First Object.** To determine the direction and extent of the present workings of a mine.

We have to measure:

1. Azimuths, or directions right and left.
2. Lengths or distances.
3. Heights, or distances up and down, either by perpendicular or angular levelling; usually, the latter.

This being done, the relative positions of all the points are known by their three rectangular coördinates.

They are referred, 1st, to a vertical plane (which may be either north and south, or pass through the first line of the survey); 2d, to another vertical plane, perpendicular to the preceding one; and 3d, to a horizontal datum-plane.

(174.) The Old Method. In the old method of Mining Surveying, a compass[1] is used for determining the azimuths. One form of the compass used for this purpose, and called a "dial," is divided from 0° to 360°. In a "right-hand dial," so called, the numbers run as on a watch-face; the 90° point being then on the east. In a "left-hand dial" they run in a contrary direction. In each, the zero-point goes ahead, and the 180°-point is at the eye.

The bearings are taken in the usual manner, a lamp being the object sighted to, instead of the rod used in work on the surface.

To test the accuracy of the bearing of a line taken at one end of it, set up the compass at the other end, or point sighted to, and look back to a lamp held at the first station, or point where the compass had been placed originally. The reading of the needle should now be the same as before.

If the position of the sights had been reversed, the reading would be the *Reverse Bearing;* a former bearing of N. 30° E. would then be S. 30° W., and so on.

If the back-sight does not agree with the first or forward sight, this latter must be taken over again. If the same difference is again found, this shows that there is *local attraction* at one of the stations; i. e., some influence, such as a mass of iron-ore, ferruginous rocks, etc., which attracts the needle, and makes it deviate from its usual direction.

To discover at which station the attraction exists, set the compass at several intermediate points in the line which joins

---

[1] For a complete description of the compass, and method of using it, the variation of the magnetic needle, and methods of determining the true meridian, see L. S. Part III.

the two stations, and take the bearing of the line at each of these points. The agreement of several of these bearings, taken at distant points, will prove their correctness.

When the difference occurs in a series of lines, proceed thus: Let C be the station at which the back-sight to B differs from the fore-sight from B to C. Since the back-sight from B to A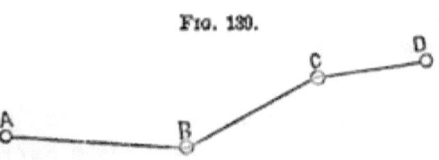

Fig. 139.

is supposed to have agreed with the fore-sight from A to B, the local attraction must be at C, and the forward bearing must be corrected by the difference just found between the fore and back sights, adding or subtracting it, according to circumstances. An easy method is to draw a figure for the case, as in Fig. 140. In it, suppose the true bearing of B C, as given by a fore-sight from B to C, to be N. 40° E., but that there is local attraction at C, so that the needle is drawn aside 10°, and points in the direction S'N', instead of SN. The back-sight from C to B will then give a bearing of N. 50° E.; a difference, or correction for the next fore-sight, of 10°. If the next fore-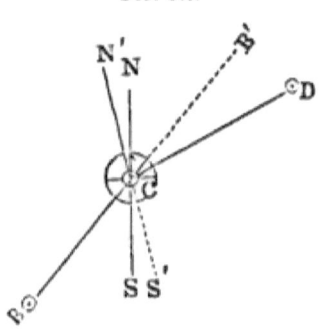

Fig. 140.

sight, from C to D, be N. 70° E., this 10° must be subtracted from it, making the true fore-sight N. 60° E.

A general rule may also be given. *When the back-sight is greater than the fore-sight*, as in this case, subtract the difference from the next fore-sight, if that course and the preceding one have both their letters the same (as in this case, both being N. and E.), or both their letters different; or add the difference if either the first or last letters of the two courses are different. *When the back-sight is less than the fore-sight*, add the difference in the case in which it has just been directed to subtract it, and subtract it where it was before directed to add it.

When the compass indicates much local **attraction**, the difference between the directions of two meeting lines (or the "*angle of deflection*" of one from the other), can still be correctly measured, by taking the difference of the bearings of the two lines, as observed at the same point. For, the error caused by the local attraction, whatever it may be, affects both bearings equally, inasmuch as a "bearing" is the angle which a line makes with the direction of the needle, and that here remains fixed in some one direction, no matter what, during the taking of the two bearings. Thus, in Fig. 140, let the true bearing of B C, i. e., the angle which it makes with the line S N, be, as before, N. 40° E., and that of C D N. 60° E. The true "angle of deflection" of these lines, or the angle B C D, is therefore 20°. Now, if local attraction at C causes the needle to point in the direction S'N', 10° to the left of its proper direction, B C will bear N. 50° E., and C D N. 70° E., and the difference of these bearings, i. e., the angle of deflection, will be the same as before.

In chaining, the leader holds a lamp in the same hand with the end of the chain, so as to be put in line. When this is done, the follower drops his end of the chain, and goes on to find the pin, or mark made by the leader, before the leader leaves it.

A gallery of a mine is thus surveyed like a road.

To measure angles of elevation or depression of the floor of the gallery, a fine string or wire is stretched parallel to the slope, and to it a divided semicircle is attached, and the angle noted by a plumb-line suspended from its centre. See Fig. 71.

(175.) The New Method. The work by the old method is very imperfect, owing to the variation of the magnetic needle, the liability of error from local attraction, and the want of precision in reading the angles, both horizontal and vertical.

A transit, or theodolite, should be used. The azimuthal and vertical angles are taken at the same time; the former on the horizontal graduated circle, and the latter on the vertical circle.

Instead of measuring the angles which each line makes with the magnetic meridian, as when the compass is used, the angles measured are those which each line makes with the preceding one, or with the first line of the survey, if the method of traversing be adopted.

The polar coördinates given by the transit are to be reduced to the three coördinate planes, to obtain the rectangular coördinates.

Very great accuracy can be obtained by using three tripods. One would be set at the first station and sighted back to from the instrument placed at the second station, and a forward sight be then taken to the third tripod, placed at the third station. The instrument would then be set on this third tripod, a back-sight taken to the tripod remaining on the second station, and a fore-sight taken to the tripod brought from the first station to the fourth station, to which the instrument is next taken; and so on. Two lamps, fitting on the tripods, are provided, to which the backward and forward sights are directed.

Owing to the irregularity of mines, and the obstacles to be overcome, great difficulties exist in mining surveying. One is that of setting up the transit. When it cannot be set upon the tripod, it is often set upon sockets which are fastened to the wall or roof of the mine.

(176.) **The Mining Transit.** In this the telescope is on one side, as shown in Fig. 141, and is balanced by a weight on the opposite side. The advantage of this form is, that sights may be taken vertically up or down, as is sometimes necessary in connecting the underground surveys with those on the surface.

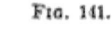

(177.) **Mapping.** The galleries of a mine on the same "level" may be platted in the same manner as a road or stream, etc.

When different "levels" are to be represented, with their connecting shafts, etc., "isometrical projection" has been used, but "military or cavalier projection" is best.

CHAPTER II.

LOCATING NEW LINES.

(178.) Second Object. To determine, on the surface of the ground, where to sink a shaft to meet a desired point in the underground workings.

To do this, repeat on the surface of the ground the survey made under it; i. e., trace on it the courses and distances of the galleries, or their equivalents. Art. (182).

The chief difficulty is to get a starting-point, and to determine the direction of the first line.

(179.) When the Mine is entered by an Adit, Fig. 142. Set the theodolite at the entrance, and get the direction of the adit,

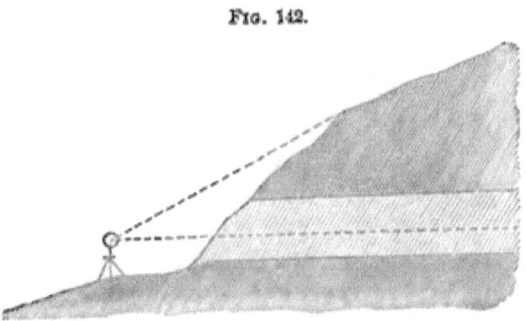

Fig. 142.

and prolong it up the hill; i. e., in the same vertical plane. The third adjustment is here important. See Art. (93).

If the line has to be prolonged by setting the instrument farther on, the second adjustment is important. Art. (93).

(180.) **When the Mine is entered by a Shaft.** Get the magnetic bearing of the first underground line, at the bottom of the shaft, with great care. Bring up the end of the line through the shaft by a plumb-line, and set the compass over this point. Set out a line with the same bearing and length as the first underground line, and repeat the succeeding courses.

WHEN THE COMPASS CANNOT BE SET OVER THE POINT, proceed thus: 1st. Find, by trial, a spot, as B (Fig. 143), which is in the correct course, and measure off a distance equal to the length of the first underground course, and then proceed as before.

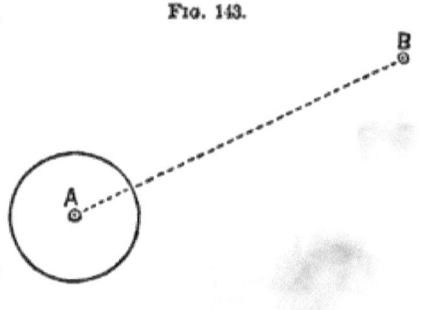

FIG. 143.

2d. *Otherwise.*—Set up anywhere, as at A′, Fig. 144, take the bearing and distance of A from A′; run a line, corresponding with the one underground, from A′ to B′. Repeat the course A′ A from B′ B; then A B is the desired line.

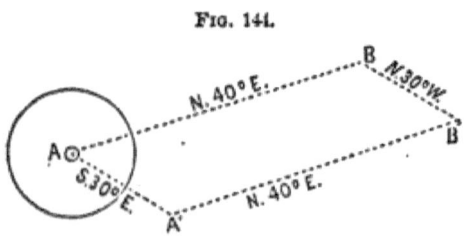

FIG. 144.

(181.) **To dispense with the Magnetic Needle.** *First Method.* Let down two plumb-lines on opposite sides of the shaft, so that their lower ends shall be very precisely in the underground line. The plumbs may be immersed in water to prevent vibration. The plumb-lines at the top of the shaft will give the required line on the surface; but its shortness is bad.

Second Method.—Set, by repeated trials, two transits on opposite sides of the shaft, so that they shall at the same time point to one another, and each, also, to one of two points in the underground line. They will then give the direction of the line above-ground.

Third Method.—If the telescope of the transit be eccentric, as in Fig. 141, set the instrument on a platform over the mouth of the shaft, so that the line of collimation of the telescope shall be in the same vertical plane with two points in the underground line, on opposite sides of the shaft. When the instrument is so placed that, in turning the telescope, the intersection of the cross-hairs strikes the two points in the underground line, the line of sight, when directed along the surface, will give the required line.

(182.) Having determined the first line, the courses of the underground survey may be repeated on the surface; or the bearing and length of a single line be calculated, which shall arrive at the desired point.

Let the zigzag line, A B, B C, C D, D Z, Fig. 145, be the courses surveyed underground, A being an adit, or at the bottom of a shaft, and Z the point to which it is desired to sink a shaft. It is required to find the direction and length of the straight line A Z.

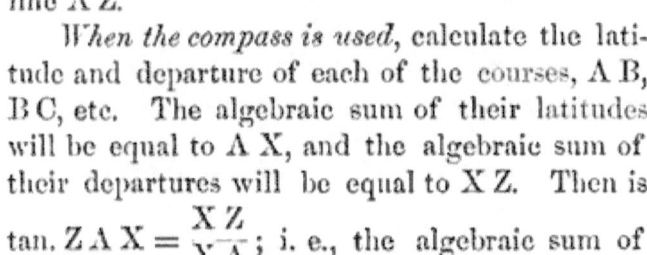

Fig. 145.

When the compass is used, calculate the latitude and departure of each of the courses, A B, B C, etc. The algebraic sum of their latitudes will be equal to A X, and the algebraic sum of their departures will be equal to X Z. Then is $\tan. Z A X = \dfrac{X Z}{X A}$; i. e., the algebraic sum of the departures divided by the algebraic sum of the latitudes is equal to the tangent of the bearing. The length of the line A Z equals the square root of the sum of the squares of A X and X Z; or equals the latitude divided by the cosine of the bearing.

When the transit is used, instead of referring all of the lines to the magnetic meridian, as in the preceding case, any line of the survey may now be taken as the meridian, as in "traversing."

In Fig. 146 all of the courses are referred to the first line of the survey. As before, a right-angled triangle will be formed. Tan. $ZAX = \dfrac{XZ}{XA}$,

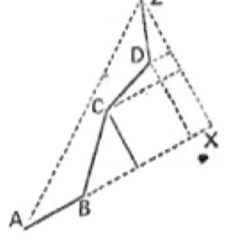

Fig. 146.

and the length of $AZ = \sqrt{AX^2 + XZ^2}$; or $AX \div \cos. XAZ$.

Two or more lines may be substituted for the single line in the two preceding cases; the condition being, that the algebraic sums of their latitudes and of their departures shall be equal to those of the underground survey.

(183.) **Third Object.** To direct the workings of a mine to any desired point.

This is the converse of the second object. We repeat under the ground the courses run above-ground; or their equivalents, as in Art. (182).

In Fig. 147, let A B, B C, C D, D Y, be the present workings of a mine, and Z the shaft to which the workings are to be directed.

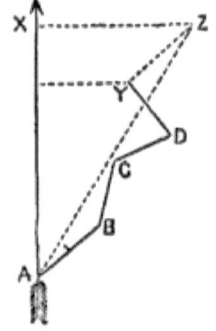

Fig. 147.

Find the latitude and departure of A Z. Then the difference between the algebraic sum of the latitudes of the underground courses already run, and the latitude of A Z, is the latitude of the required course; and the difference between the algebraic sum of the departures of the underground lines, and the departure of A Z, is the departure of the required course.

The length of Y Z equals the square root of the sum of the squares of its latitude and departure.

(184.) **Problems.** Most of the problems which arise in Mining Surveying can be solved by an application of the familiar principles of geometry and trigonometry.

1. Given, the angle which a vein makes with the horizon,

and the place where it meets the surface, to find how deep a shaft at D will be required to strike the vein:

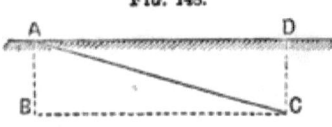

Fig. 148.

$$DC = AD . \tan . DAC.$$

2. Given, the depth of the shaft DC, and the "dip" of the vein, to find where it crops out:

$$AD = DC . \cot . DAC.$$

3. Given, the depth of a shaft when the vein "crops out," and the "dip" of the vein, to find the distance from the bottom of the shaft to the vein:

$$BC = AB . \cot . ACB.$$

If the ground makes an angle with the horizon, then the problems involve oblique-angled triangles instead of right-angled triangles, as in the preceding cases. Their solution, however, is quite as simple.

In the more difficult problems, the measurement of lines is required, one or both ends of which are inaccessible. For a full investigation of this subject, see "Gillespie's Land Surveying," Part VII.

PART VI.

THE SEXTANT, AND OTHER REFLECTING INSTRUMENTS.

CHAPTER I.

THE INSTRUMENTS.

(185.) Principle. The angle subtended at the eye by lines passing from it to two distant objects, may be measured by so arranging two mirrors that one object is looked at directly, and the other object is seen by its image, reflected from one mirror to the second, and from the second mirror to the eye. If the first mirror be moved so that the doubly-reflected image of the second object be made to cover or coincide with the object seen directly, then is the desired angle equal to twice the angle which the mirrors make with each other.

PROOF.—Let two mirrors be parallel. Then a ray of light, striking one of them, reflected to the other, and reflected again from that, would pass off in a direction parallel to its first direction.

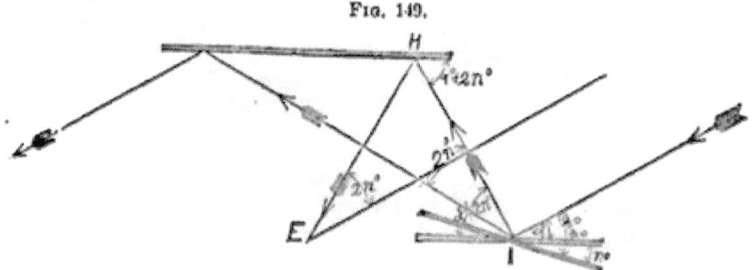

FIG. 149.

Let a equal the angle between the incident ray and the

first mirror. Now, suppose the first mirror to be turned $n°$. The incident ray now makes an angle with this mirror $n°$ greater than before; it will therefore pass off, making an angle with the mirror $n°$ greater than before. But the mirror itself now makes an angle of $n°$ with its former direction; therefore, the ray will pass off at an angle of $a° + 2n°$ with the former surface of the mirror, and in a direction differing $2n°$ from its former direction, and the direction of the ray reflected from the second mirror will therefore differ $2n°$ from its former direction.

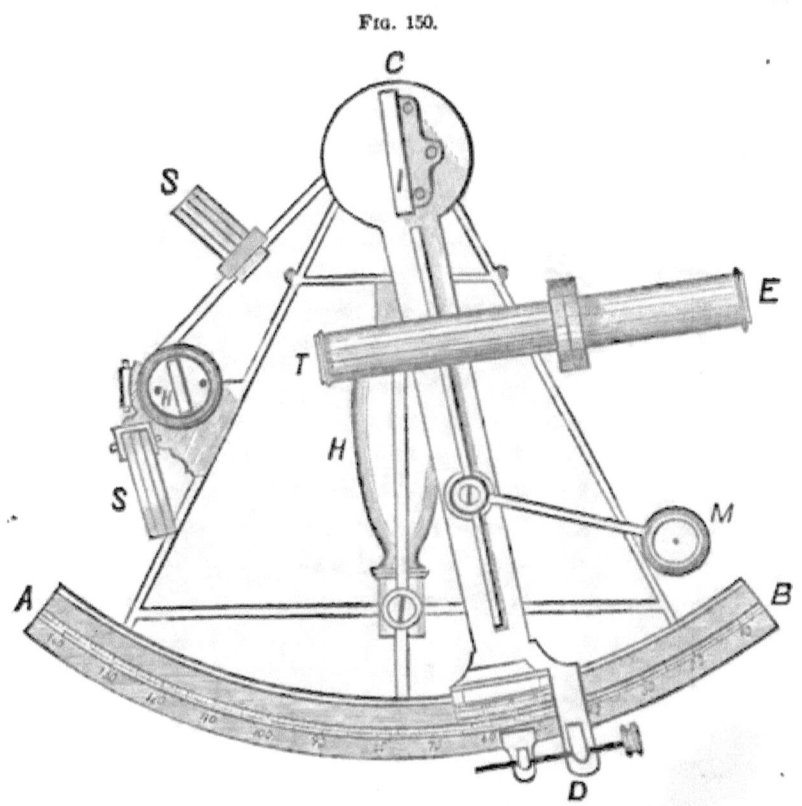

Fig. 150.

If, now, an eye, placed at E, sees an object in the second mirror, in the direction E H, which has been reflected from two mirrors, then the line E H makes an angle with the true direction of the line equal to twice the angle which the mirrors make with one another. If the eye also sees an object, directly in the line E II, which apparently coincides with the reflected image of the first object, then is the angle, subtended at the eye by the lines passing to it from the two objects, equal to twice the angle which the two mirrors make with one another.

(186.) **Description of the Sextant,** Fig. 150. The frame is usually of brass, constructed so as to combine strength with lightness. The handle, H, by which it is held, is of wood. A B is a graduated arc; C D, the index-arm, is movable about a pivot in the centre of the graduated arc. M is a glass, which may be moved over the vernier to aid in reading it. The index-glass, I, is a small mirror, attached to the index-arm, so as to be perpendicular to the plane of the graduated arc. The

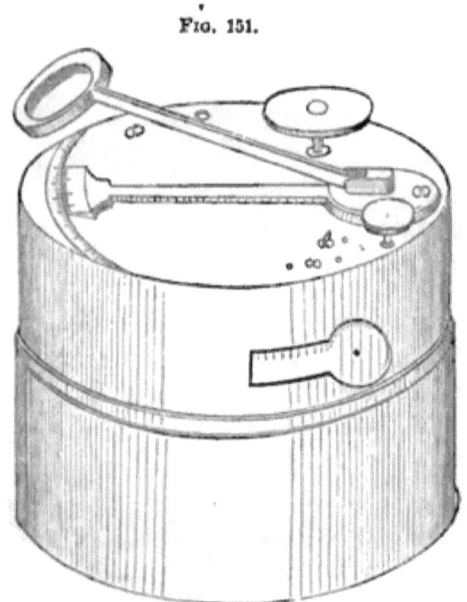

Fig. 151.

horizon-glass, H, is attached perpendicularly to the plane of the instrument, and parallel to the index-glass when the index is at zero. The lower half of this glass is silvered, to make it a reflector, and the upper half is transparent. T E is the telescope; S S are sets of colored glasses, used to moderate the light of the sun, when that body is observed.

The sextant has an arc of one-sixth of a circle, and measures angles up to 120°, the divisions of the graduated arc being numbered with twice their real value, so that the true desired angle, subtended by the two objects, is read off at once. The arc is usually graduated to 10' and subdivided by a vernier to 10".

(187.) The box or pocket sextant, shown in Fig. 151, has the same glasses as the larger sextant, enclosed in a circular box, about three inches in diameter. The lower part, which answers for a handle when in use, screws off and is used for a cover, making the instrument only half as deep as it appears in the figure.

The octant has an arc of one-eighth of a circumference, and measures angles to 90°.

(188.) **The Reflecting Circle.** This is an instrument constructed on the same principle, and used for the same purposes, as the sextant. In it the graduated arc extends to the whole circumference, and more than one vernier may be used by producing the index-arm to meet the circumference in one or two more points.

(189.) **Adjustments of the Sextant.** 1. *To make the index-glass perpendicular to the plane of the arc:*

Bring the index near the centre of the arc, and place the eye near the index-glass, and nearly in the plane of the arc. See if the part of the arc reflected in the mirror appears to be a continuation of the part seen directly. If so, the glass is perpendicular to the plane of the arc. If not, adjust it by the screws behind it.

2. *To make the horizon-glass perpendicular to the plane of the arc:*

The index-glass having been adjusted, sight to some well-defined object, as a star, and if, in moving the index-arm, one image seems to separate from or overlap the other, then the horizon-glass is not perpendicular to the plane of the arc. It must be made so by the screws attached to it.

Another method of testing the perpendicularity of the horizon-glass is as follows: Hold the instrument vertically, and bring the direct and reflected images of a smooth portion of the distant horizon into coincidence. Then turn the instrument until it makes an angle with the vertical. If the two images still coincide, the glasses are parallel; and, as the index-glass has been made perpendicular to the plane of the arc, the horizon-glass is in adjustment.

3. *To make the line of collimation of the telescope parallel to the plane of the arc:*

The line of collimation of the telescope is an imaginary line, passing through the optical centre of the object-lens, and a point midway between the two parallel wires. These wires are made parallel to the plane of the sextant by revolving the tube in which they are placed.

To see whether the line of collimation of the telescope is in adjustment, bring the images of two objects, such as the sun and moon, into contact at the wire nearest the instrument, and then, by moving the instrument, bring them to the other wire. If the contact remains perfect, the line of collimation is parallel to the plane of the arc; if it does not, the adjustment must be made by the screws in the collar of the telescope.

4. *To see if the two mirrors are parallel when the index is at zero:*

Bring the direct and reflected images of a star into coincidence. If the index is at zero, then no correction is necessary; if not, the reading is the "*index-error*," and is positive or negative, according as the index is to the right or left of the zero.

The "index-error" may be rectified by moving the horizon-glass until the images do coincide when the index is at zero, but it is usually merely noted, and used as a correction, being added to each reading if the error is positive, or subtracted from each reading if the error is negative.

(190.) How to observe. Hold the instrument so that its plane is in the plane of the two objects to be observed, and hold it loosely. Look through the eye-hole, or plain tube, or telescope, at the left hand or lower object, by direct vision, through the unsilvered part of the horizon-glass. Then move the index-arm till the other object is seen in the silvered part of the horizon-glass, and the two are brought to apparently coincide. Then the reading of the vernier is the angle desired.

If one object be brighter than the other, look at the former by reflection. If the brighter objects be to the left or below, hold the instrument upside down.

If the angular distance of the object be more than the range of the sextant (about 120°), observe from one of them to some intermediate object, and thence to the other.

A good rest for a sextant is an ordinary telescope-clamp, through which is passed a stick, one end of which is fitted into a hole made in the sextant-handle, and the other end of which is weighted for a counterpoise.

FIG. 152.

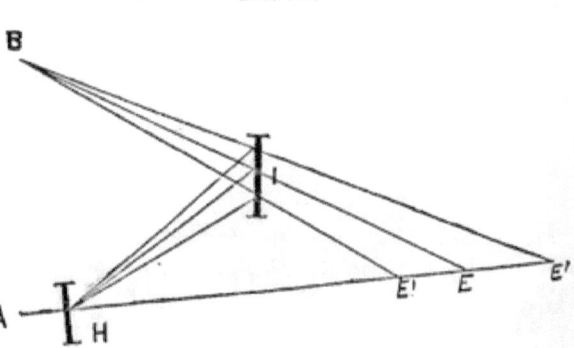

(191.) Parallax of the Sextant. The angle observed with the sextant is that made by two lines: one, B I, passing from the reflected object to the index-glass, and which is thence reflected to the horizon-glass, and thence to the eye; and the other, H E, passing from the object directly to the eye; i. e., the angle which B I produced makes with H E. But the eye may be at E' on either side of E. Then we require the angle which B E′ makes with H E. These angles are the same only when the eye is in the same line with B I produced; i. e., when E′ coincided with E. In all other cases, the observed angle differs from the desired angle by the small angle E B E′, which is called the *parallax* of the instrument.

It is the angle which would be subtended at the reflected object B by the distance E E′. It is usually very small for distant objects. Thus, at a mile's distance, 1 inch subtends an angle of only 3 seconds, and of 3 minutes at 100 feet distance.

To escape it, if one object be distant and the other near, view the former by reflection. If both be near, find some distant point in line with one of them, and view this new point by reflection, and the other near one directly.

CHAPTER II.

THE PRACTICE.

(192.) **To set out Perpendiculars.** Set the index at 90°. Hold the instrument over the given point by a plumb-line, and look along the line by direct vision. Send a rod in about the desired direction, and when it is seen by reflection to coincide with the point on the line looked at directly, it will be in a line perpendicular to the given line at the desired point.

Conversely, to find where a perpendicular from a given point would strike a line:

Set the index at 90°, and walk along the line, looking directly at a point on it, until the given point is seen by reflection to coincide with the point on the line. A plumb-line let fall from the eye will give the desired point.

(193.) **The Optical Square**, Fig. 153. This is a box containing two mirrors, fixed at an angle of 45° to each other, and therefore giving an angle of 90°, as does the sextant with its glasses fixed at that angle. It is used to set out perpendiculars.

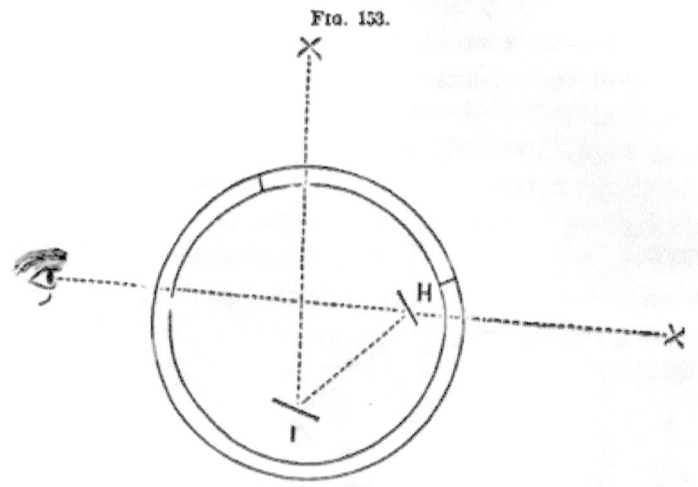

Fig. 153.

(194.) **To measure a Line, one End being inaccessible.** Let A B be the required line, and B the inaccessible point.

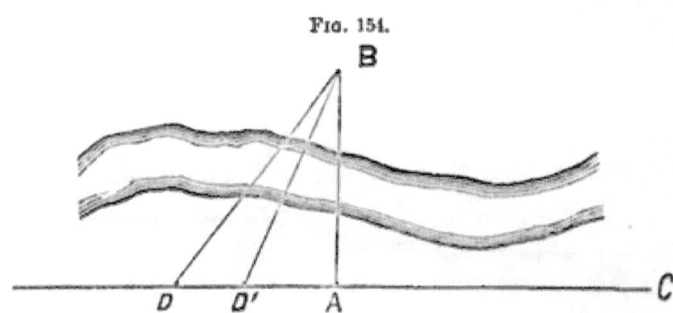

Fig. 154.

At A set off a perpendicular, A C, by Art. (192). Then set the index at 45°, and walk backward from A in the line C A prolonged, looking by direct vision at C, until you arrive at some point, D, from which B is seen by reflection to coincide with C. Then is A D = A B.

If more convenient, after setting off the right angle, set the index at 63° 26′, and then proceed as before. The objects will be seen to coincide when at some point D′. Then A D′ = ½ A B. If the index be set at 71° 34′, then the measured distance will be ⅓ A B, and so on.

If the index be set at the complements of the above angles, the distance measured will be, in the first case, twice, and in the second case three times the desired one.

When the distance A D cannot be measured, as in Fig. 155, fix D as before. Set the index at 26° 34′, and go along

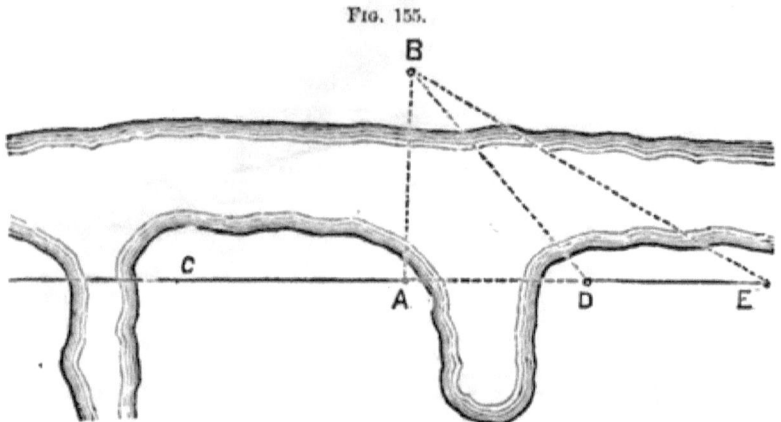

Fig. 155.

the line to E, where the objects are seen to coincide with each other; then is A E twice A B, and hence E D = A B.

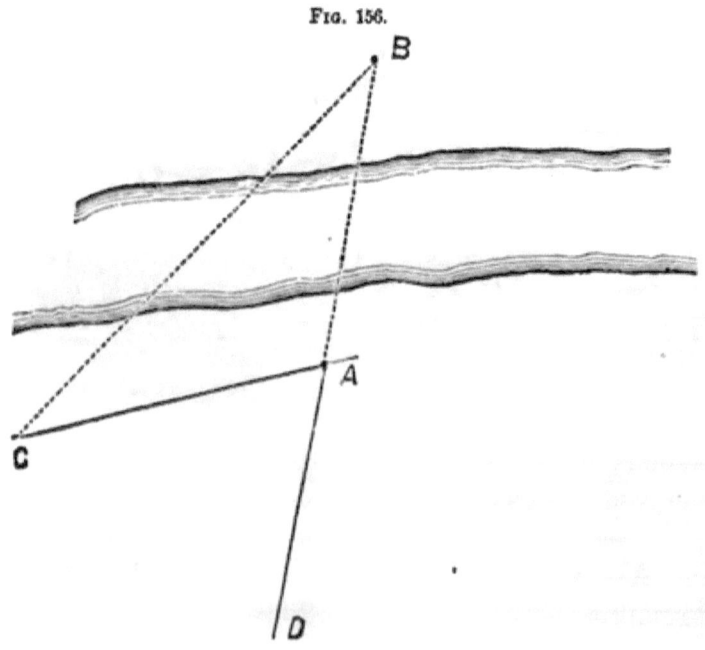

Fig. 156.

(195.) *Otherwise.* At A set off an angle, as C A D (A D being a prolongation of A B). Then walk along the line A C with the index set to half that angle, looking at A directly, and B by reflection, till you come to some point, C, at which they coincide. Then is C A = A B.

(196.) **To measure a Line when both Ends are inaccessible.** Let A B be the required line. At any point, C, measure

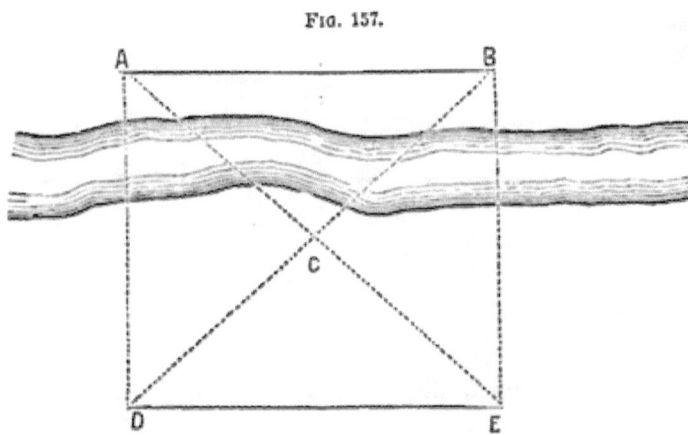

Fig. 157.

the angle A C B. Set the sextant to half that angle, and walk back in the line B C prolonged till at some point, D, A and B are seen to coincide, as in last problem; thus making A C = C D. Do the same on A C produced to some point, E. Then is D E = A B.

(197.) All the methods for overcoming obstacles to measurement, determining inaccessible distances, etc. (L. S. Part VII.), with the transit or theodolite, can be executed with the sextant.

(198.) **To measure Heights.** Measure the vertical angle between the top of the object and a mark at the height of the eye, as with a theodolite or transit, and then calculate the height as in Part II., Art. (98).

Otherwise. Set the index at 45°, and walk backward till the mark and the top of the object are brought to coincide. Then the horizontal *distance* equals the *height*.

So, too, if the index is set at 63° 26′, the height equals twice the distance, and so on. The ground is supposed to be level.

Fig. 138.

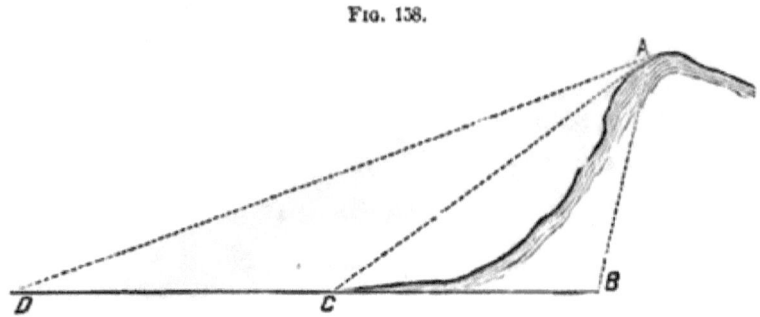

When the Base is inaccessible: Make C = 45°, and D = 26° 34′. Then C D = A B. So, too, if C = 26° 34′, and D = 18° 26′.

This may be used when a river flows along the base of a hill whose height is desired, or in any other like circumstance.

(199.) To observe Altitudes in an artificial Horizon. In this

Fig. 139.

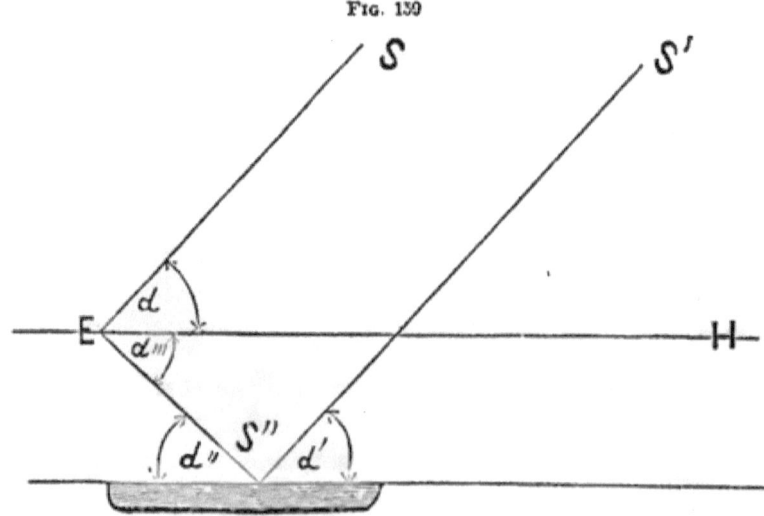

method we measure the angle subtended at the eye between the object and its image reflected from an artificial horizon of mercury, molasses, oil, or water. The image of the object in the mercury is looked at directly, and the object itself is viewed by reflection. The object observed is supposed to be so distant that the rays from it, which strike respectively the index-glass and the artificial horizon, are parallel; i. e., S and S', Fig. 159, are the same point.

Then will the observed angle SES'' be double the required angle SEH.

Demonstration.

$a = a'$, $a' = a''$, and $a'' = a'''$. Hence $a''' = a$.

$SES'' = a + a''' = 2a = 2\,SEH$.

(200.) When the sun is the object observed, to determine whether it is his upper or lower limb whose altitude has been observed, proceed thus:

Having brought two limbs to touch, push the index-arm from you. If one image passes over the other, so that the other two limbs come together, then you had the lower limb at first. If they separate, you had the upper limb.

In the forenoon, with an inverting telescope, the lower limbs are parting, and the upper limbs are approaching; and *vice versa* in the afternoon.

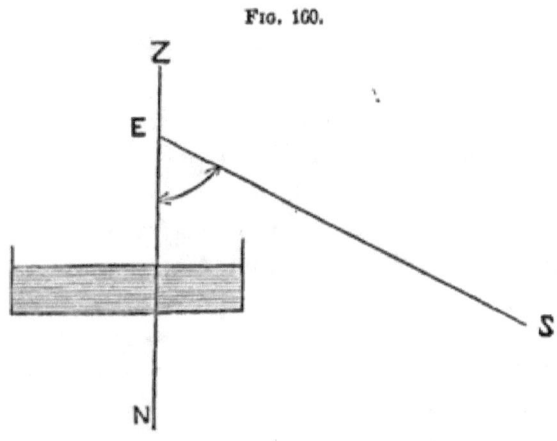

Fig. 160.

(201.) To observe very small altitudes and depressions with the artificial horizon:

Stretch a string over the artificial horizon. Place your head so that you see the string cover its image in the mercury. Then the eye and string determine a vertical plane.

Then observe, looking at the string by direct vision, and seeing the object by reflection, and you have the angle SEN, in Fig. 160, the supplement of the zenith distance.

Otherwise. Fix behind the horizon-glass a piece of white paper with a small hole in it, and with a black line on it perpendicular to the plane of the arc.

Then look into the mercury, so as to see in it the image of the line. Your line of sight is then vertical, and the angle to the object seen by reflection is measured as before.

(202.) To measure Slopes with the Sextant and Artificial Horizon. Let AB be the surface of the ground, and AF a

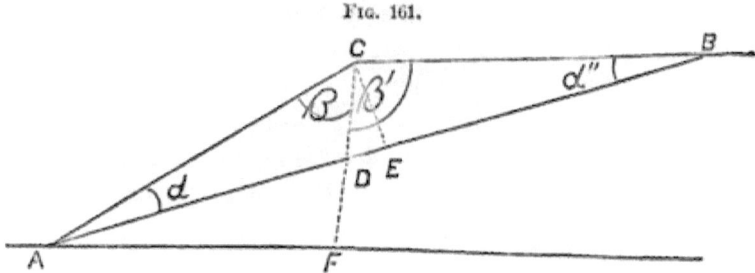

Fig. 161.

horizontal line. Mark two points equally distant from the eye. Measure, by the preceding method, the angles β and β', which CA and CB make with the vertical CD. Then will half the difference of these angles equal the angle which the slope makes with the horizon.

Demonstration. Continue the vertical line CD to meet the horizontal line in F, and draw CE perpendicular to AB. Then the triangles CDE and ADF are similar, being right-angled and having the acute angles at D equal. Consequently, the angle $DCE = DAF$, which is the angle of the slope with the horizon. But $DCE = \frac{1}{2}(\beta' - \beta)$, hence $\frac{1}{2}(\beta' - \beta) = $ the angle which the slope of the ground makes with the horizon.

If the points A and B be not equally distant from C, but yet far apart, this method will still give a very near approximation, the error, which is additive, being $\frac{1}{2}(a' - a)$.

Demonstration.

$$D C E = \beta' + a' - 90°,$$
$$D C E = -\beta - a + 90°,$$
$$\overline{2 D C E = \beta' - \beta + a' - a,}$$
$$D C E = \tfrac{1}{2}(\beta' - \beta) + \tfrac{1}{2}(a' - a).$$

(203.) **Oblique Angles.** When the plane of two objects, observed by the sextant, is very oblique to the horizon, the observed angle will differ much from the horizontal angle which is its horizontal projection, and which is the angle needed for platting. The projected angle may be larger or smaller than the observed angle.

This difficulty may be obviated in various ways:

1. Observe the angular distance of each object from some third object, very far to the right or left of both. The difference of these angles will be nearly equal the desired angle.

2. Note, if possible, some point above or below one of the objects, and on the same level with the other, and observe to it and the other object.

3. Suspend two plumb-lines, and place the eye so that these lines cover the two objects. Then observe the horizontal angle between the plumb-lines.

4. For perfect precision, observe the oblique angle itself, and also the angle of elevation or depression of each of the objects. With these data the oblique angle can be reduced to its horizontal projection, either by descriptive geometry or more precisely by calculation, thus:

Let A H B be the observed angle, and A' H B' the required horizontal angle.

Conceive a vertical H Z, and a spherical surface, of which H, the vertex of the angle, is the centre. Then will the vertical planes, A H A' and B H B', and the oblique plane A H B, cut this sphere in arcs of great circles, Z A", Z B", and A" B",

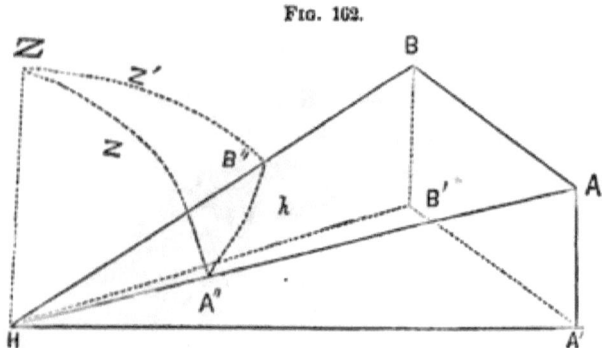

Fig. 162.

thus forming a spherical triangle, A" Z B", in which A" B" = h measures the observed angle; Z A" = Z measures the zenith distance of the point A; and Z B" = Z' measures the zenith distance of the point B.

These zenith distances are observed directly, or given by the observed angles of elevation or depression. Then we have the three sides of the triangle to find the angle Z = A' H B'.

Calling P the half sum of the three sides we have:

$$\text{Sin. } \tfrac{1}{2} Z = \sqrt{\frac{\sin. (P - Z) \sin. (P - Z')}{\sin. Z . \sin. Z'}}.$$

An approximate correction, when the zenith distances do not differ from 90° by more than 2° or 3°, is this:

$$\left(90° - \frac{Z+Z'}{2}\right)^2 \tang. \tfrac{1}{2} h . \sin. 1" - \left(\frac{Z-Z'}{2}\right)^2 \cot. \tfrac{1}{2} h . \sin. 1".$$

The quantities in the parentheses are to be taken in seconds.

The answer is in seconds, and additive.

(204.) The advantages of the sextant over the theodolite are these:

---

[1] See Jackson's Trigonometry, page 65, Fifth Case.

1. It does not require a fixed support, but can be used while the observer is on horseback, or on a surface in motion, as at sea.

2. It can take simultaneous observations on two moving bodies, as the moon and a star.

It can also do all that the theodolite can. Its only defect is in observing oblique angles in some cases. By these properties it determines distances, heights, time, latitude, longitude, and true meridian, and thus is a portable observatory.

PART VII.

MARITIME OR HYDROGRAPHICAL SURVEYING.

(**205.**) The object of this is to fix the positions of the deep and shallow points in harbors, rivers, etc., and thus to discover and record the shoals, rocks, channels, and other important features of the locality.

The relative positions of prominent points on the shore are very precisely determined by "Trigonometrical Surveying," Part VIII. These form the basis of operations, and afford the means of correcting the results obtained by the less accurate methods employed for filling in the details.

CHAPTER I.

THE SHORE LINE.

(**206.**) **The High-water Line.** The principal points on the high-water line are determined by triangulating, Art. (233). The sections between these points are surveyed with the compass and chain; by running a series of straight lines so as to follow, approximately, the shore line, and taking offsets from the straight lines of the survey to the bends in the shore line. The straight lines can be more accurately determined by "traversing" with the transit. Art. (94).

(207.) **The Low Water-Line.** In "tidal-waters" this is more difficult, because low and bare for only a short time. The survey is best made with the sextant, observing from prominent points to three signals, by the trilinear method—Art. (213)—and sketching by the eye bends of the shore between the stations observed from.

There should be one to observe and one to record. Let 1 and 2, Fig. 163, be two points on the low-water line, whose position it is desired to determine. The observations taken will be as follows:

FIG. 162.

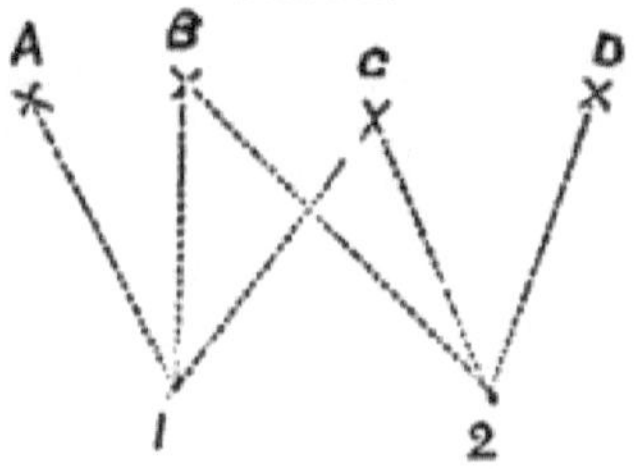

(1.) A and B . . . 18°
    B and C . . . 20°

(2.) B and C . . . 15°
    C and D . . . 45°

When the shore is inaccessible, a base line must be measured on the water, and points on the shore fixed by angles from its ends, as in Art. (232).

(208.) **Measuring the Base.** 1. By sound. Sound travels at the rate of 1,090 feet per second, with the temperature at 30° Fahr. For higher or lower temperatures, add or subtract 1¼ foot for each degree. If the wind blows with or against the movement of the sound, its velocity must be added or subtracted. If it blows obliquely, the correction will be its velocity multiplied by the cosine of the angle which the direction of the wind makes with the direction of the sound.

2. By measuring with the sextant the angular height of the mast of a vessel, then we have:

Distance = height of mast ÷ tan. of the angle.

SOUNDINGS.   133

3. By astronomical observations at points 50 miles apart, more or less, determining their latitudes and longitudes, and hence knowing their distance and bearings. A vessel may be

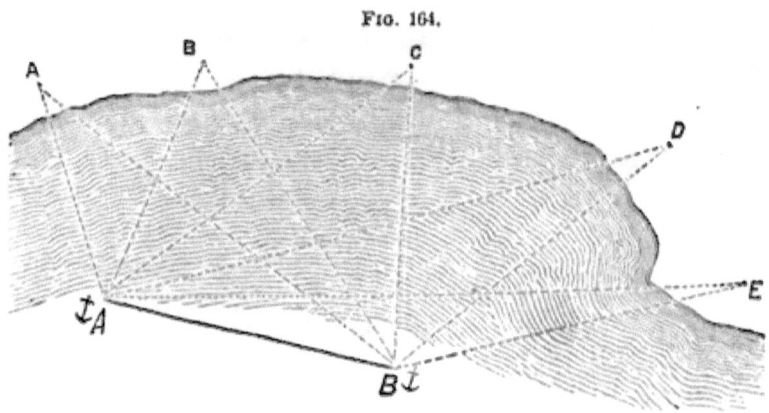

Fig. 164.

anchored at various points between A′ and B′, and thus new base lines be formed, from the ends of which to triangulate to points on the shore.

CHAPTER II.

SOUNDINGS.

(209.) In a river or narrow water, the soundings may be taken in zigzag lines, from shore to shore, at equal intervals of time, as in Fig. 165.

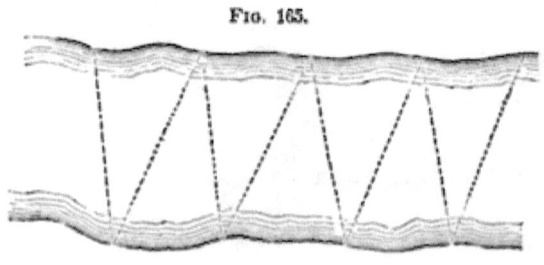

Fig. 165.

134   LEVELLING, TOPOGRAPHY, AND HIGHER SURVEYING.

(210.) **On a Sea-coast.** The position of the boat in the water, when the soundings are taken, must be determined at regular intervals. This is done in various ways.

(211.) **From the Shore.** By observing with a compass or transit to the boat from stations on the shore, at a given signal or fixed time.

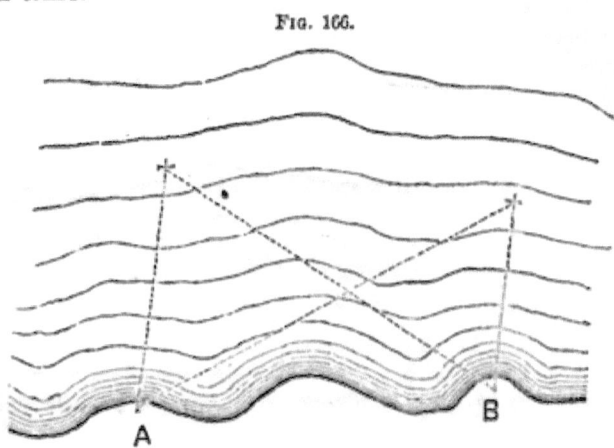

Fig. 166.

Its place is then fixed as in Art. (232). Two observers are necessary. Three are better, as the third checks the other two. This is accurate in theory, but not in practice, simultaneous observations being impracticable, and confusion frequent. Also, more men are required.

(212.) **From the Boat with a Compass.** Establish signals along the shore, Art. (240), distinguish them by colors, or by the number of cross-pieces on the staff, thus: + ± +, and observe to them from the boat with a prismatic compass (L. S. 232), or Burnier's compass, Art. (96). The place of the boat is then determined, and may be fixed on the map by drawing, from the two known points, lines having the *opposite* bearings, and their intersection will be the required point. This is rapid and easy, but not precise.

(213.) **From the Boat with the Sextant.** This is the trilinear method, and is the best. Two observers, or two sextants with one observer, are necessary.

(**214.**) TRILINEAR SURVEYING is founded on the method of determining the position of a point by measuring the angles between three lines conceived to pass from the required point to three known points. Thus, in the figure, the point P is determined by the angles A P B and B P C, the points A, B, and C, being known. To fix the place of the point from these data is known as the "problem of the three points." It

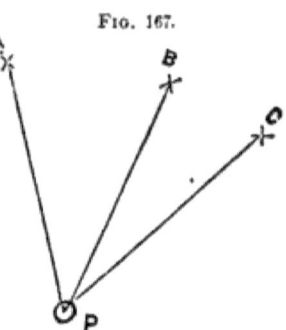

Fig. 167.

will be here solved geometrically, instrumentally, and analytically.

(**215.**) **Geometrical Solution.** Let A, B, and C, be the known objects observed from S, the angles A S B and B S C being there measured. To fix this point, S, on the plat containing

Fig. 168.

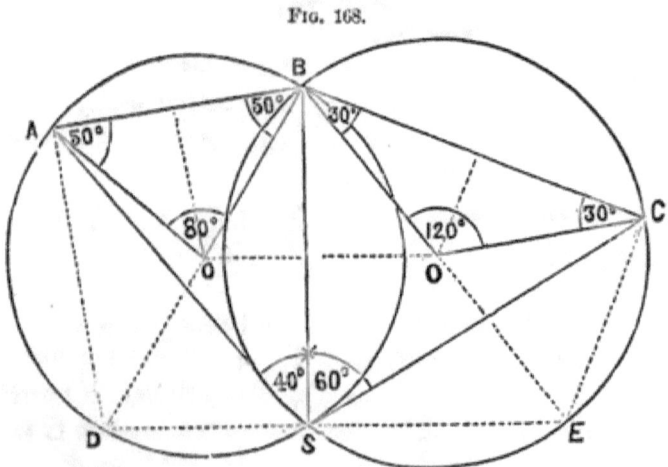

A, B, and C, draw lines from A and B, making angles with A B, each equal to $90° - $ A S B. The intersection of these lines at O will be the centre of a circle passing through A and B, in the circumference of which the point S will be situated.[1]

[1] For, the arc A B measures the angle A O B at the centre, which angle $= 180° - 2(90° - $ A S B$) = 2$ A S B. Therefore, any angle inscribed in the circumference and measured by the same arc is equal to A S B.

Describe this circle. Also, draw lines from B and C, making angles with BC, each equal to 90° − BSC. Their intersection, O′, will be the centre of a circle passing through B and C. The point S will lie somewhere in its circumference, and, therefore, in its intersection with the former circumference. The point is thus determined:

In the figure the observed angles, ASB and BSC, are supposed to have been respectively 40° and 60°. The angles set off are therefore 50° and 30°. The central angles are consequently 80° and 120°, twice the observed angles.

The dotted lines refer to the checks explained in the latter part of this article.

When one of the angles is obtuse, set off its difference from 90° on the opposite side of the line joining the two objects to that on which the point of observation lies.

When the angle ABC is equal to the supplement of the sum of the observed angles, the position of the point will be indeterminate, for the two centres obtained will coincide, and the circle described from this common centre will pass through the three points, and any point of the circumference will fulfil the conditions of the problem.

A third angle, between one of the three points and a fourth point, should always be observed, if possible, and used like the others, to serve as a check.

Many tests of the correctness of the position of the point determined may be employed. The simplest one is, that the centres of the circles, O and O′, should lie in the perpendiculars drawn through the middle points of the lines AB and BC.

Another is, that the line BS should be bisected perpendicularly by the line OO′.

A third check is obtained by drawing at A and C perpendiculars to AB and CB, and producing them to meet BO and BO′, produced in D and E. The line DE should pass through S; for, the angles BSD and BSE being right angles, the lines DS and SE form one straight line.

The figure shows these three checks by its dotted lines.

(216.) **Instrumental Solution.** The preceding process is tedious where many stations are to be determined. They can be more readily found by an instrument called a *Station-pointer*, or *Chorograph*. It consists of three arms, or straight-edges, turning about a common centre, and capable of being set so as to make with each other any angles desired. This is effected by means of graduated arcs carried on their ends, or by taking off with their points (as with a pair of dividers) the proper distance from a scale of chords constructed to a radius of their length. Being thus set so as to make the two observed angles, the instrument is laid on a map containing the three given points, and is turned about till the three edges pass through these points. Then their centre is at the place of the station, for the three points there subtend on the paper the angles observed in the field.

A simple and useful substitute is a piece of transparent paper, or ground glass, on which three lines may be drawn at the proper angles and moved about on the paper as before.

(217.) **Analytical Solution.** The distances of the required point from each of the known points may be obtained analytically. Let $AB = c$; $BC = a$; $ABC = B$; $ASB = S$; $BSC = S'$. Also, make $T = 360° - S - S' - B$. Let $BAS = U$; $BCS = V$. Then we shall have:

$$\cot U = \cot T \left(\frac{c \cdot \sin S'}{a \cdot \sin S \cdot \cos T} + 1 \right).$$

$$V = T - U.$$

$$SB = \frac{c \cdot \sin U}{\sin S}; \text{ or, } = \frac{a \cdot \sin V}{\sin S'}.$$

$$SA = \frac{c \cdot \sin ABS}{\sin S}. \qquad SC = \frac{a \cdot \sin CBS}{\sin S'}.$$

Attention must be given to the algebraic signs of the trigonometrical functions.

Example. $ASB = 33° 45'$; $BSC = 22° 30'$; $AB = 600$ feet; $BC = 400$ feet; $AC = 800$ feet. Required the distances and directions of the point S from each of the stations.

In the triangle ABC, the three sides being known, the angle ABC is found to be $104° 28' 39''$. The formula then gives the angle $BAS = U = 105° 8' 10''$; whence BCS is found to be $94° 8' 11''$; and $SB = 1042.51$; $SA = 710.193$; and $SC = 934.291$.

(218.) Between Stations. Positions of the boat are thus observed only at considerable distances apart, and the boat is rowed from one of these points to a second one, and soundings taken at regular intervals of time between them.

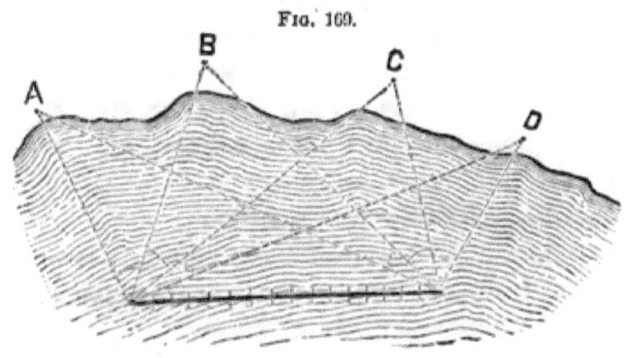

FIG. 169.

The distance apart of the soundings depends on the regularity of the bottom, the depth of the water, and the object of the survey. Care should be taken to leave no spot unexplored.

For great accuracy, anchor at some point, and determine its place as above, and then proceed to another point, paying out a line, fastened to the anchor, and sounding at regular distances. Cast anchor at the second point, go back to the first, take up the anchor, go on to the second, and then proceed as before.

The soundings, or depth of the water, are made with rods, chains, or lines, according to the depth and the precision required.

(**219.**) The sounding-line should be strong and pliable. The lead fastened to its extremity should be shaped like the frustum of a cone. The size of the line and the weight of the lead will depend upon the depth of the water. The hand lead-line is limited to twenty fathoms and is marked thus: at three fathoms a piece of leather; at five, a white rag; at seven, a red rag; at ten, a piece of leather with a round hole in it; at thirteen, a blue rag; at fifteen, a white rag; at seventeen, a red rag; at twenty, a piece of cord with two knots. These divisions are called *marks*. The other divisions called *deeps* are: at half a fathom a piece of leather with three points; at one fathom a piece of leather with one point; at two fathoms a piece of leather with two points; at four, six, eight, eleven, fourteen, sixteen, and eighteen fathoms a piece of cord with a knot in it; at nine and twelve, a piece of cord with two knots in it. The deep-sea lead-line is similarly marked up to twenty fathoms, and each additional fathom is indicated by a cord with an additional knot, and half-way between these a piece of leather marks the five fathoms.

FIG. 170.

The length of the line should be frequently tested.

The character of the bottom is determined by placing tallow into a hollow in the base of the lead, which adheres to the material at the bottom. A barbed pike is sometimes attached to the base of the lead for this purpose.[1] Fig. 170.

---

[1] For description of sounding apparatus, see U. S. C. S. Report, 1860; for deep-sea soundings, Reports of 1854, 1858, and 1859.

CHAPTER III.

TIDE-WATERS.

(**220.**) The soundings taken must all be reduced to mean low spring-tides.

The tides are semidiurnal oscillations of the ocean, caused by the combined attractions of the moon and sun, especially the former. The tide is not a current, but a broad, flat wave. The lunar one follows the moon, about 30° east of it. The solar one follows the sun. In the open sea, it is high water about 30° east of the moon. To the east of this the tide is ebbing; west of this it is rising; 90° distant it is low tide.

The tides, being caused by the attraction of the sun and moon, are greatest when they act together; i. e., at new and full moon, or a day or two after. These are *spring-tides*.

They are least when the sun and moon act perpendicularly to each other. These are *neap-tides*. The spring-tides are the highest and lowest tides. The moon makes a tidal wave which recurs in 12 hours 24 minutes, while the sun's recurs in 12 hours. The coincidence of these waves produces spring-tides, and their partial cancellation neap-tides.

(**221.**) On the Atlantic coast, the successive high waters and successive low waters are nearly at equal heights above or below the mean, with intervals of 12 hours 24 minutes.

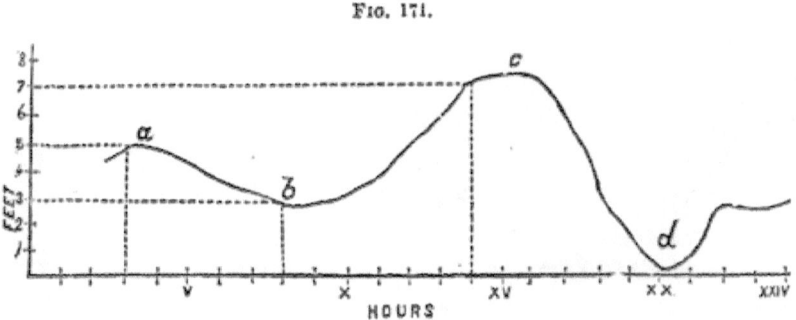

Fig. 171.

On the Pacific coast, the successive high and low tides may differ several feet in height, and several hours in intervals.

In the preceding diagram, we see this: a and c are high tides, and b and d are low tides.

(**222.**) The mean level of the sea is the mean between the mean of two successive high tides and the intermediate low tide. This is constant, while high and low tides vary.

(**223.**) Care should be taken not to confound the time of high water or low water with the time of the turn of the tide. The flood-tide may run for hours after the time of high water, and so, too, the stream of the ebb after low water. "Slack water" is the interval of time during which there is no current.

(**224.**) The "Establishment" of a place is a standard tide-time, i. e., the number of hours at which high water occurs after the moon's transit over the meridian. This varies with the age of the moon. When observed on the days of full or change, it is the "Vulgar Establishment of the Port." The "Corrected Establishment" is the mean of the intervals between the times of transit and times of high tide for half a month. This is used for finding the time of high water on any given day, and tables are constructed from observations at the principal ports for finding the correction for semi-monthly inequality. On the Atlantic coast of the United States this inequality is about three-quarters of an hour.

(**225.**) **Tide-Gauges.** The simplest form is a stick of timber, graduated to feet and inches, or tenths, and either set up in the water, or fastened to the face of a dock, or pier, so that the rise of the tide may be noted on it. The zero-point of each gauge is referred to a permanent "bench-mark" on the shore.

On account of the difficulty of sustaining a timber of considerable height against the force of the wind and waves, several successive gauges are sometimes used, the bottom line of each gauge higher up being on a level with the top line of the next lower one.

Such an arrangement is required on gentle slopes. On the sea-coast, a tube is put outside of a float, thus rendering the surface inside more tranquil. Tide-gauges should be self-registering.[1]

(226.) **Tide-Tables.** Observations of tides are recorded thus:

STATION.	Interval between time of moon's transit (southing) and time of high water.		RISE AND FALL.			MEAN DURATION OF		
	Mean.	Difference between greatest and least	Mean.	Spring.	Neap.	Flood.	Ebb.	Stand.
	H. M.	H. M.	Feet.	Feet.	Feet.	H. M.	H. M.	H. M.
Portland,	11 25	0 44	8.8	10.0	7.6	6 14	6 12	0 20
Boston,	11 22	0 44	10.1	13.1	7.4	6 16	6 18	0 09
New York,	8 13	0 46	4.3	5.4	3.4	6 00	6 25	0 28
Charlestown,	7 13	0 36	5.3	6.3	4.6	6 36	6 09	0 33
Key West,	9 22	1 12	1.4	2.3	0.7	6 59	5 25	0 12
San Francisco,	12 03	1 22	3.9	5.0	2.9	6 30	5 52	0 30

[1] For a description of a self-registering tide-gauge, see U. S. C. S. Report, 1853, and for a sea-coast tide-gauge, the Report of 1854.

TIDE WATERS. 143

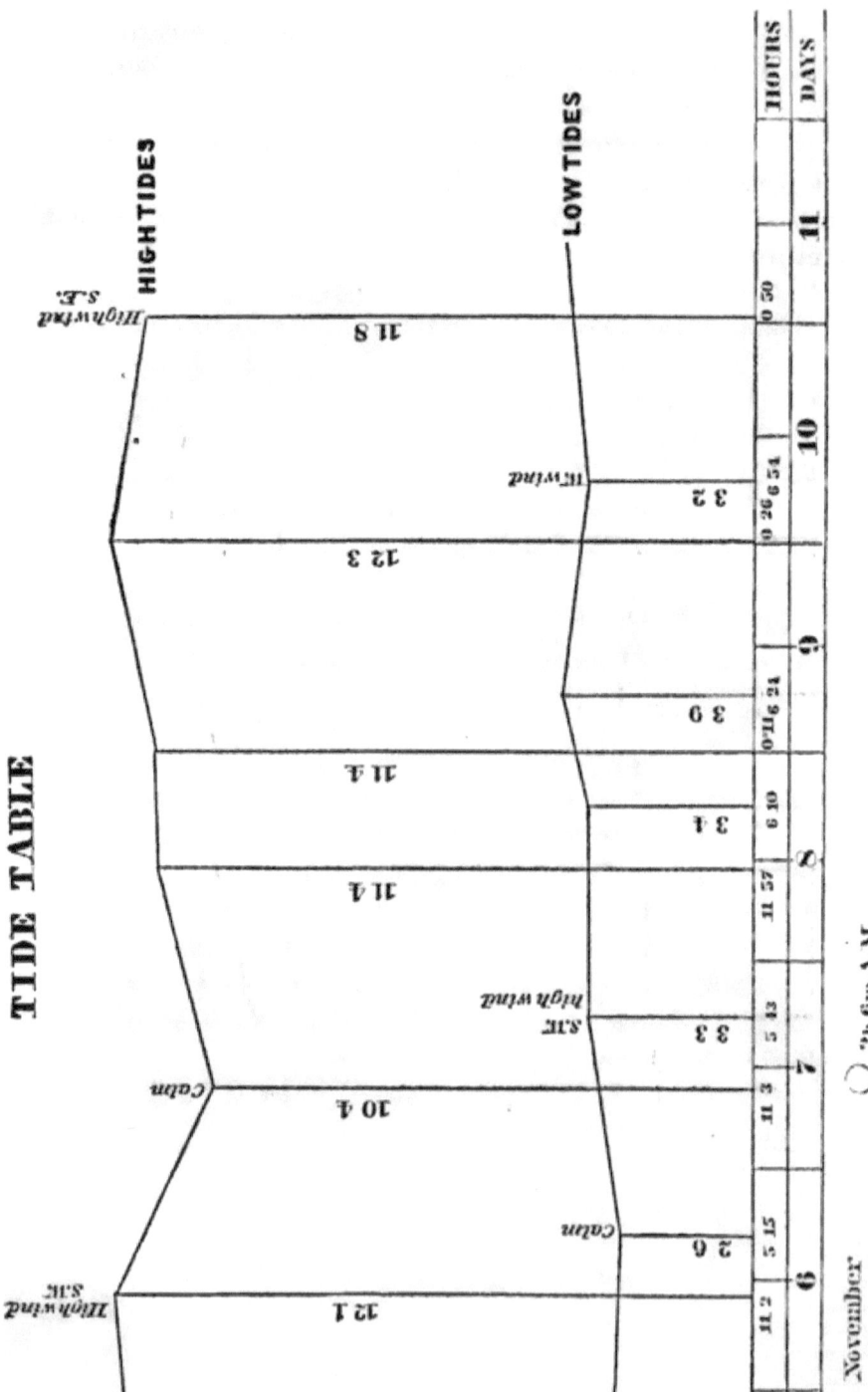

The horizontal line of the abscissas is divided into months, days, and half days. The hours and minutes of the heights of the tides are noted at the feet of the vertical ordinates, in black for high water, and red for low water. The lines of the ordinates are black for high water, and red for low water, and are on a scale of one inch to 4.4 feet ($=\frac{1}{.52.8}=100$ feet to 1 mile). The heights of the tides are noted in feet and tenths at the side of the ordinates, and the weather at their summits.

(227.) In rivers, a number of tide-gauges are necessary, at moderate distances apart, especially at the bends, because the tidal lines of high and low water are not parallel to one another.

The soundings are to be reduced by the nearest gauge, or by the mean of the two between which they may be taken.

(228.) **Beacons and Buoys.** Beacons are permanent objects, such as piles of stones with signals on them, usually on shoals and dangerous rocks.

Buoys are floating objects, such as barrels, or hollow iron spheres or cylinders, anchored by a chain, and variously painted, to indicate either dangers or channels.

Those placed by the United States Coast Survey are so colored and numbered that in entering a bay, harbor, or channel, red buoys with even numbers shall be passed on the starboard or right hand, black buoys with odd numbers on the port hand or left hand, and buoys with red and black stripes, on either hand. Buoys in channel-ways are colored with alternate white and black vertical stripes.

On dangerous coasts, self-ringing bells and "fog-whistles" are used.

CHAPTER VI.

THE CHART.

(229.) Having determined the lines of high and low water, the position of the channels, rocks, shoals, etc., and the soundings, a chart must be made, on which all these are laid down in their proper places. For scales see Art. (171.)

The high-water line is platted like the bounding lines of a farm. The points determined in the low-water line, and the positions of the boat, determined by the method given in Art. (213), are fixed on the chart by one of the methods given in Arts. (215), (216), and (217). Contour curves are drawn as in land topography (Part IV.), for the first four fathoms. These

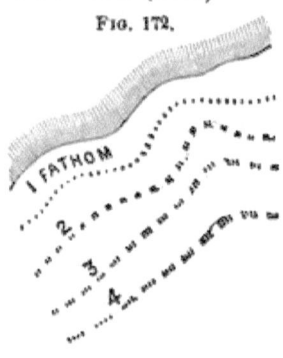

Fig. 172.

may be indicated by dotted lines, as in Fig. 172, or they may be shaded with Indian-ink, as in Fig. 173.

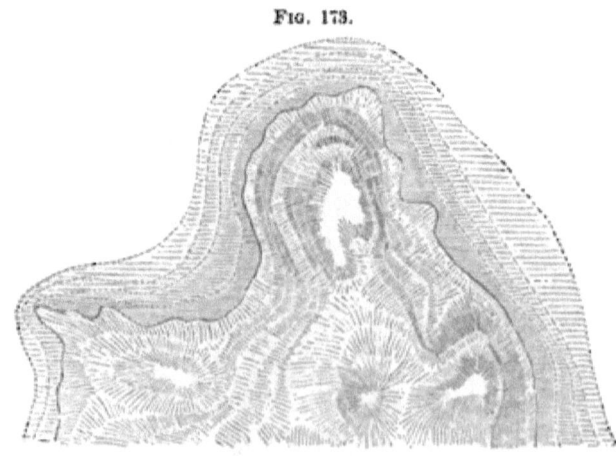

Fig. 173.

Beyond four fathoms, the depths are noted in fathoms and vulgar fractions.

146   LEVELLING, TOPOGRAPHY, AND HIGHER SURVEYING.

(230.) Various conventional signs are used; some of the principal ones are given in Figs. 174–194:

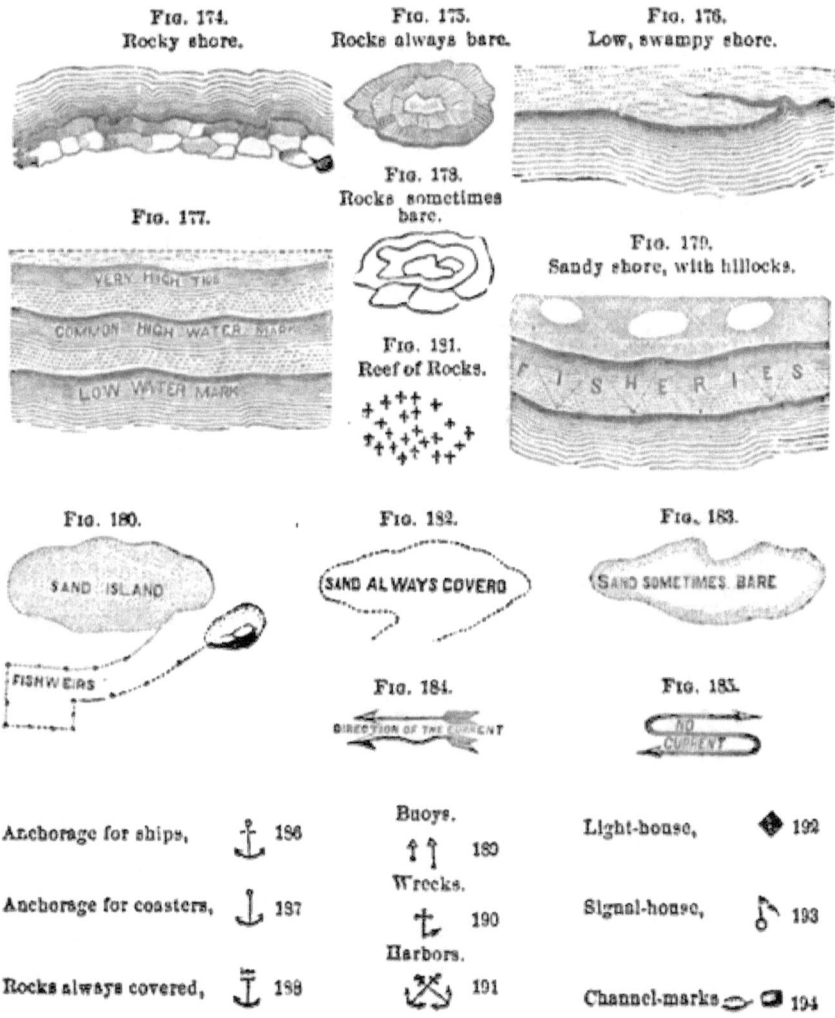

PART VIII.

SPHERICAL SURVEYING, OR GEODESY.

CHAPTER I.

THE FIELD-WORK.

(231.) Nature. It comprises the methods of surveying surfaces of such extent that the curvature of the earth cannot be neglected. The method of triangulation is usually employed.

(232.) TRIANGULAR SURVEYING is founded on the method of determining the position of a point by the intersection of two known lines. Thus, the point P is determined by knowing the length of the line A B, and the angles P B A and P A B, which the lines P A and P B make with A B. By an extension of the principle, a field, a farm, or a country, can be surveyed by measuring

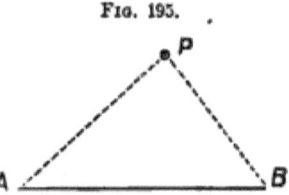

Fig. 195.

only one line, and calculating all the other desired distances, which are made sides of a connected series of imaginary *Triangles*, whose angles are carefully measured. The district surveyed is covered with a sort of net-work of such triangles, whence the name given to this kind of surveying. It is more commonly called "Trigonometrical Surveying," and sometimes "Geodesic Surveying," but improperly, since it does not necessarily take into account the curvature of the earth, though always adopted in the great surveys in which that is considered.

(233.) Outline of Operations. A *base-line*, as long as possible (five or ten miles in surveys of countries), is measured with extreme accuracy.

From its extremities, angles are taken to the most distant objects visible, such as steeples, signals on mountain-tops, etc.

The distances to these and between these are then calculated by the rules of trigonometry.

The instrument is then placed at each of these new stations, and angles are taken from them to still more distant stations, the calculated lines being used as new base-lines.

This process is repeated and extended till the whole district is embraced by these "primary triangles" of as large sides as possible.

One side of the last triangle is so located that its length can be obtained by measurement as well as by calculation, and the agreement of the two proves the accuracy of the whole work.

Within these primary triangles, *secondary* or smaller triangles are formed, to fix the position of the minor local details, and to serve as starting-points for common surveys with chain and compass, etc. Tertiary triangles may also be required.

The larger triangles are first formed, and the smaller ones based on them, in accordance with the important principle in all surveying operations, always to work from the whole to the parts, and from greater to less.

(234.) Measuring a Base. Extreme accuracy in this is necessary, because any error in it will be *multiplied* in the subsequent work. The ground on which it is located must be smooth and nearly level, and its extremities must be in sight of the chief points in the neighborhood. Its point of beginning must be marked by a stone set in the ground with a bolt let into it. Over this a theodolite or transit is to be set, and the line "ranged out." The measurement may be made with chains (which should be formed like that of a watch), etc., but best with rods. We will notice, in turn, their *materials, supports, alignment, levelling,* and *contact.*

As to *materials*, iron, brass, and other metals, have been used, but are greatly lengthened and shortened by changes of temperature. Wood is affected by moisture. Glass rods and tubes are preferable on both these accounts. But wood is the most convenient. Wooden rods should be straight-grained white pine, etc., well seasoned, baked, soaked in boiling oil, painted and varnished. They may be trussed, or framed like a mason's plumb-line level, to prevent their bending. Ten or fifteen feet is a convenient length. Three are required, which may be of different colors, to prevent mistakes in recording. They must be very carefully compared with a standard measure.

Supports must be provided for the rods, in accurate work. Posts, set in line at distances equal to the length of the rods, may be driven or sawed to a uniform line, and the rods laid on them, either directly or on beams a little shorter. Tripods, or trestles, with screws in their tops to raise or lower the ends of the rods resting on them, or blocks with three long screws passing through them and serving as legs, may also be used. Staves, or legs, for the rods have been used, these legs bearing pieces which can slide up and down them, and on which the rods themselves rest.

The *alignement* of the rods can be effected, if they are laid on the ground, by strings, two or three hundred feet long, stretched between the stakes set in the line, a notched peg being driven when the measurement has reached the end of one string, which is then taken on to the next pair of stakes; or, if the rods rest on supports, by projecting points on the rods being aligned by the instrument.

The *levelling* of the rods can be performed with a common mason's level; or their angle measured, if not horizontal, by a "slope-level."

The *contacts* of the rods may be effected by bringing them end to end. The third rod must be applied to the second before the first has been removed, to detect any movement. The ends must be protected by metal, and should be rounded (with radius equal to length of rod), so as to touch in only one point. Round-headed nails will answer tolerably. Better are small steel cylinders, horizontal on one end and vertical on the other. Sliding ends, with verniers, have been used. If one rod be higher than the next one, one must be brought to touch a plumb-line which touches the other, and its thickness be added. To prevent a shock from contact, the rods may be brought not quite in contact, and a wedge be let down between them till it touches both at known points on its graduated edges. The rods may be laid side by side, and lines drawn across the end of each be made to coincide or form one line. This is more accurate. Still better is a "visual contact," a double microscope with cross-hairs being used, so placed that one tube bisects a dot at the end of one rod, and the other tube bisects a dot at the end of the next rod. The rods thus never touch. The distance between the two sets of cross-hairs is of course to be added.

A base could be measured over very uneven ground, or even water, by suspending a series of rods from a stretched rope by rings in which they can move, and levelling them and bringing them into contact as above.

The most perfect base-measuring apparatus is that used on the United States Coast Survey.[1] It consists of a bar of brass and a bar of iron, a little less than six metres long, supported parallel to each other, firmly attached to a block at one end, and left free to move at the other, so that the entire contraction and expansion are at that end. At right angles to these bars is a short lever, called the "lever of compensation." It is attached to the lower (brass) bar at the free end by a hinge, and an agate knife-edge on the lever rests against a steel plate at the end of the iron bar.

When the temperature is raised, both bars expand, but the brass one more than the iron one, so that the upper end of the lever of compensation is thrown back. A knife-edge, turned outward, is placed on the lever, at such a distance from the other knife-edge and the hinge, that it shall remain unmoved by equal changes of temperature in the two bars.

Brass and iron, exposed to the same temperature, will not

[1] For a full description, see Coast Survey Report of 1854.

heat equally in equal times. To overcome this difficulty, the bars are given equal absorbing surfaces, but their cross-sections are adapted to their different specific heats and conducting powers.

The knife-edge on the upper end of the lever of compensation presses against a short sliding rod, supported on the upper (iron) bar, and held firmly against the lever by a spiral spring. The sliding rod is terminated on the outer end by an agate plane.

The end of the apparatus we have been considering is called the compensating end. We will now consider the sector end, where are arranged the parts for adjusting the contacts between the successive rods in measuring; and for determining the inclination of the rod on sloping ground.

This end also terminates in a sliding rod, bearing on its extremity an agate knife-edge, placed horizontally, and resting by its inner end against an upright "lever of contact." This lever is fastened by a hinge at the lower end, and its upper end rests against a tongue, attached to the "level of contact," which is mounted on trunnions. When the sliding rod is moved in, the lever of contact presses against the tongue of the level of contact and turns the level. The inner end of the level-tube is weighted so as to insure a constant pressure when the contact is made between two rods, and the bubble is brought to the centre. The sector is an arrangement for determining the angle at which the rod is inclined.

The whole apparatus is enclosed in a double tin tubular case, only the ends of the sliding rods, bearing the agates, being exposed. The observations are taken through glass doors in the side of the tube. The extreme length is six metres. Two of these tubes are used in measuring a base, and each are supported by two trestles. The tubes are aligned by the aid of a transit.

On one base, seven miles long, measured with this apparatus, the greatest supposable error was computed, from remeasurements, to be less than six-tenths of an inch. On another base, six and three-quarter miles long, the probable error was less than one-tenth of an inch, and the greatest supposable error was less than three-tenths of an inch.

(235.) Corrections of Base. If the rods were not level, their length must be reduced to its horizontal projection. This would be the square root of the difference of the squares of the length of the rod (or of the base), and of the height of one end above the other; or the product of the same length by the cosine of the angle which it makes with the horizon.[1]

If the rods were metallic, they would need to be corrected for temperature. Thus, if an iron bar expands $\frac{7}{1000000}$ of its length for 1° Fahr., and had been tested at 32°, and a base had been measured at 72° with such a bar 10 feet long, and found to contain 3,000 of them, its apparent length would be 30,000 feet, but its real length would be 8.4 feet more.

EXPANSION FOR 1° FAHRENHEIT.

Brass bar = 0.00001050903;
Iron bar = 0.000006963535;
Platinum = 0.0000051344;
Glass = 0.0000043119;
White Pine = 0.0000022685.

(236.) Reducing the Base to the Level of the Sea. Let $AB = a$ be the measured base, and $A'B' = x$, the base reduced to the level of the sea, h the height of the measured base above the level of the sea, and r the radius of the earth to the level of the sea. Then we have:

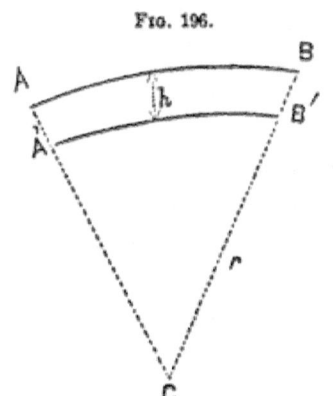

Fig. 196.

$$r + h : r :: a : x$$

$$\therefore x = a\frac{r}{r+h}$$

$$a - x = \frac{ah}{r+h} = \frac{\frac{ah}{r}}{1+\frac{h}{r}} = \frac{ah}{r}\left(1 + \frac{h}{r}\right)^{-1}$$

---

[1] More precisely, A being this angle, and not more than 2° or 3°, the difference between the inclined and horizontal lengths equals the inclined or real length multiplied by the square of the minutes in A, and that by the decimal 0.00000004231.

Developing by the binomial formula, we get:

$$a - x = a\frac{h}{r} - a\frac{h^2}{r^2} + a\frac{h^3}{r^3} -, \text{etc.}$$

As h is very small in comparison with r, the first term of the correction is generally sufficient.

(237.) **A Broken Base.** When the angle C is very obtuse, the lines A C and C B being measured, and forming nearly a

Fig. 197.

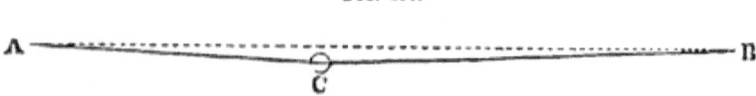

straight line, the length of the line A B is found thus: Naming the lines, as is usual in trigonometry, by small letters corresponding to the capital letters at the angles to which they are opposite, and letting K = the number of minutes in the supplement of the angle C, we shall have:

$$AB = c = a + b - 0.0000000042308 \times \frac{abK^2}{a+b}.$$

Log. $0.0000000042308 = 2.6264222 - 10.$

(238.) **Base of Verification.** As mentioned in Art. (233), a side of the last triangle is so located that it can be measured, as was the first base. If the measured and calculated lengths agree, this proves the accuracy of all the previous work of measurement and calculation, since the whole is a chain of which this is the last link, and any error in any previous part would affect the very last line, except by some improbable compensation. How near the agreement should be, will depend on the nicety desired and attained in the previous operations. Two bases, 60 miles distant, differed on one great English survey 28 inches; on another, 1 inch; and on a French triangulation extending over 500 miles, the difference was less than 2 feet. Results of equal or greater accuracy are obtained on the United States Coast Survey. "The Fire Island base, on the south side of Long Island, and the Kent Island base in Chesapeake Bay, are connected by a primary triangulation. This Kent Island base is 5 miles and 4 tenths long, and the original Fire Island base is 8 miles and 7 tenths. The shortest distance between them is 208 miles, but the distance through the triangulation is 320. The number of intervening triangles is 32, yet the computed and measured lengths of the Kent Island base exhibit a discrepancy no greater than 4 inches."

(239.) **Choice of Stations.** The stations, or "trigonometrical points," which are to form the vertices of the triangles, and to be observed to and from, must be so selected that the resulting triangles may be "well-conditioned," i. e., may have such sides and angles that a small error in any of the measured quantities will cause the least possible errors in the quantities calculated from them. The higher calculus shows that the triangles should be as nearly equilateral as possible. This is seldom attainable, but no angle should be admitted less than 30°, or more than 120°. When two angles only are observed, as is often the case in the secondary triangulation, the unobserved angle ought to be nearly a right angle.

To extend the triangulation, by continually increasing the sides of the triangles, without introducing "ill-conditioned" triangles, may be effected as in Fig. 198. A B is the measured base, C and D are the nearest stations. In the triangles A B C and A B D, all the angles being observed, and the side A B known, the other sides can be readily calculated. Then, in each of the triangles D A C and D B C, two sides and the contained angles are given to find D C, one calculation checking the other. D C then becomes a base to calculate E F, which is then used to find G H, and so on.

The fewer primary stations used the better, both to prevent confusion and because the smaller number of triangles makes the correctness of the results more "probable."

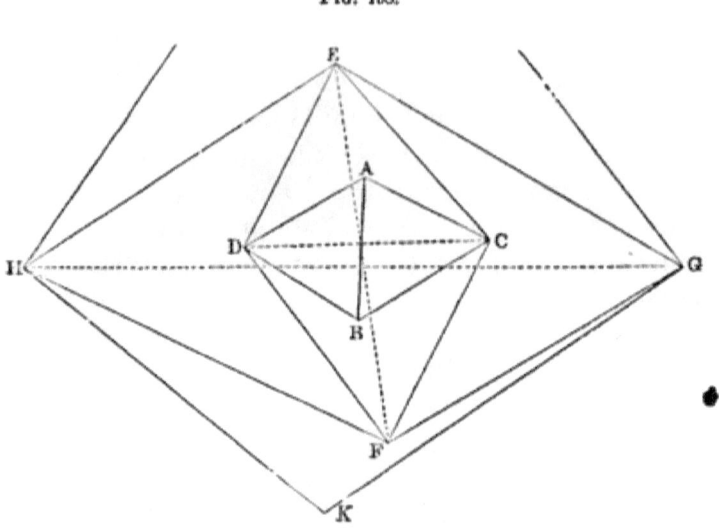

Fig. 108.

The United States Coast Survey, under the superintendence of Prof. A. D. Bache, displays some fine illustrations of these principles, and of the modifications they may undergo to suit various localities. The figure on the opposite page represents part of the scheme of the primary triangulation resting on the Massachusetts base, and including some remarkably well-conditioned triangles, as well as the system of quadrilaterals, which is a valuable feature of the scheme when the sides of the triangles are extended to considerable lengths, and quadrilaterals, with both diagonals determined, take the place of simple triangles.

The engraving is on a scale of 1 : 1,200,000.

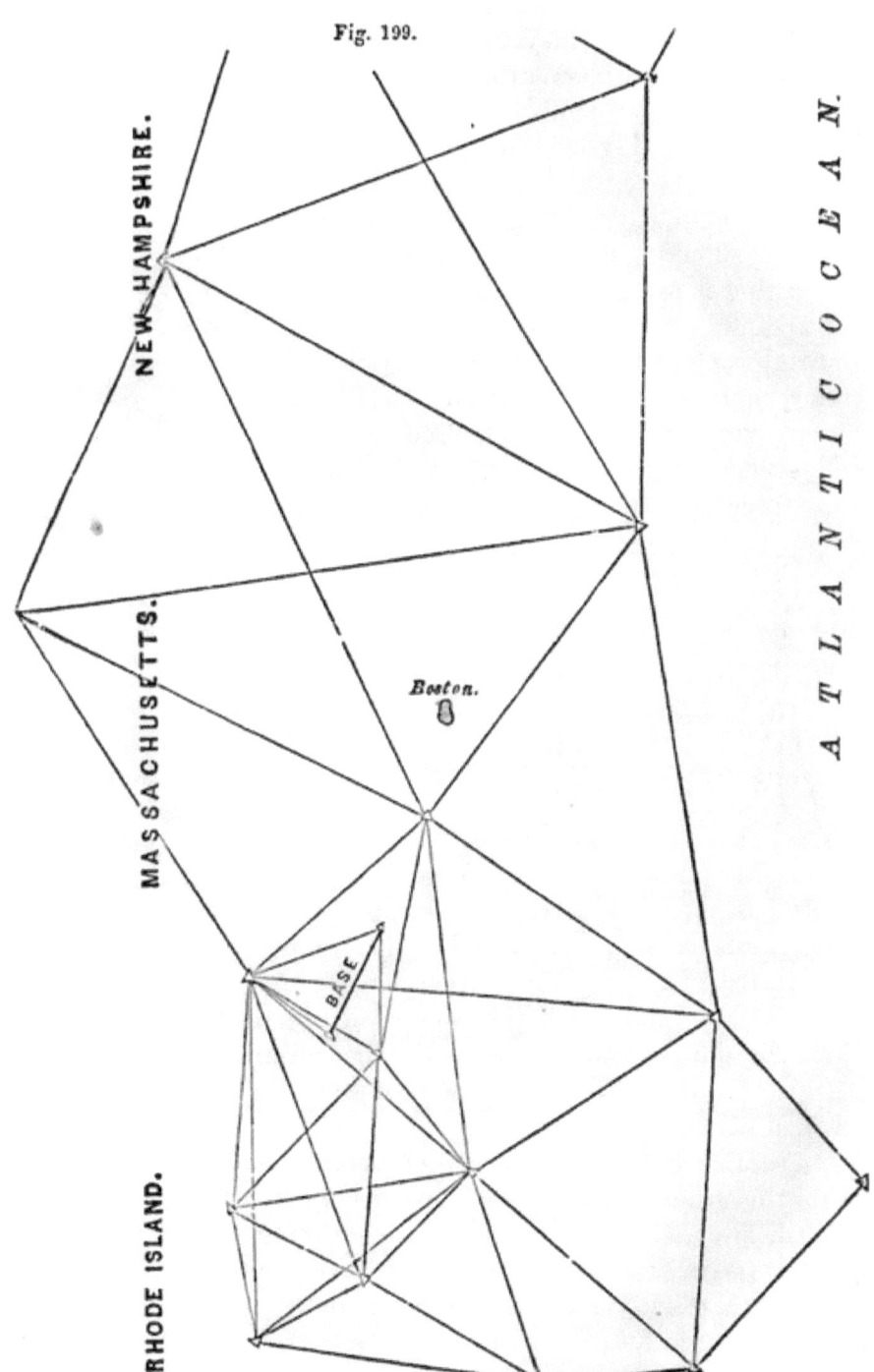

Fig. 199.

(240.) **Signals.** They must be high, conspicuous, and so made that the instrument can be placed precisely under them.

Three or four timbers framed into a pyramid, as in Fig. 200, with a long mast projecting above, fulfil the first and last conditions. The mast may be made vertical by directing two theodolites to it, and adjusting it so that their telescopes follow it up and down, their lines of sight being at right angles to each other. Guy-ropes may be used to keep it vertical.

A very excellent signal, used on the Massachusetts State Survey, by Mr. Borden, is represented in the three following figures. It consists merely of three stout

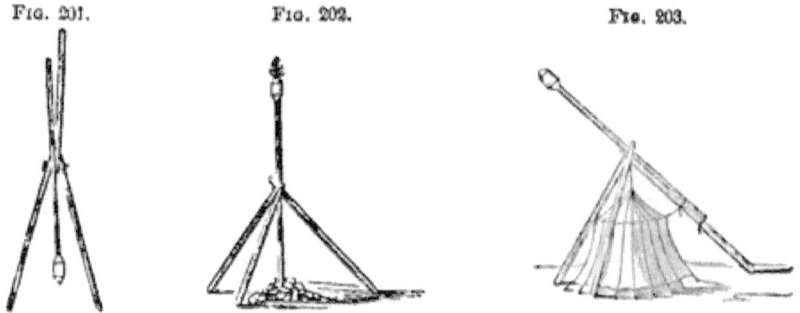

sticks, which form a tripod, framed with the signal-staff, by a bolt passing through their ends and its middle. Fig. 201 represents the signal as framed on the ground; Fig. 202 shows it erected and ready for observation, its base being steadied with stones; and Fig. 203 shows it with the staff turned aside, to make room for the theodolite and its protecting tent. The heights of these signals varied between 15 and 80 feet.

Another good signal consists of a stout post let into the ground, with a mast fastened to it by a bolt below and a collar above. By opening the collar, the mast can be turned down and the theodolite set exactly under the former summit of the signal, i. e., in its vertical axis.

A tripod of gas-pipe has been used to support the signal in positions exposed to the sea, as on shoals. It is taken to the desired spot in pieces, and there screwed together and set up.

Signals should have a height equal to at least $\frac{1}{1000}$ of their distance, so as to subtend an angle of half a minute, which experience has shown to be the least allowable.

To make the tops of the signal-masts conspicuous, flags may be attached to them; white and red, if to be seen against the ground, and red and green if to be seen against the sky.[1] The motion of flags renders them visible, when much larger motionless objects are not. But they are useless in calm weather. A disk of sheet-iron, with a hole in it, is very conspicuous. It should be arranged so as to be turned to face each station. A barrel, formed of muslin sewed together, four or five feet long, with two hoops in it two feet apart, and its loose ends sewed to the signal-staff, which passes through it, is a cheap and good arrangement. A tuft of pine-boughs fastened to the top of the staff, will be well seen against the sky.

In sunshine a number of pieces of tin, nailed to the staff at different angles, will be very conspicuous. A truncated cone of burnished tin will reflect the sun's rays to the eye in almost every situation.

The most perfect arrangement is the "heliotrope," invented by Gauss. This consists of a mirror a few inches square, so mounted on a telescope, near the eye-end, that the reflection of the sun may be thrown in any desired direction. They have been observed on at a distance of 80 or 90 miles, when the outlines of the mountains on which they were placed were invisible. A man, called a "heliotroper," is stationed

---

[1] To determine at a station A, whether its signal can be seen from B, projected against the sky or not, measure the vertical angles B A Z and Z A C. If their sum equals or exceeds 180°, A will be thus seen from B. If not, the signal at A must be raised till this sum equals 180°.

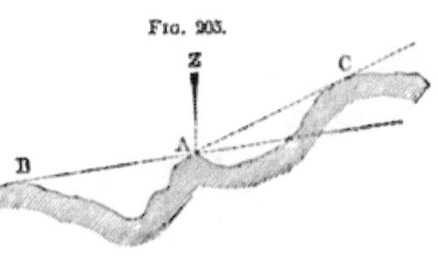

Fig. 205.

at the instrument. He directs the telescope toward the station at which the transit is placed for observation, and keeps the mirror turned so as to reflect the sun in a direction parallel to the axis of the instrument. This he accomplishes by causing the reflection to pass through two perforated disks, mounted on the telescope, one near the object-end, and the other near the mirror.

For night-signals, an Argand lamp is used; or, best of all, Drummond's light, produced by a stream of oxygen gas directed through a flame of alcohol upon a ball of lime. Its distinctness is exceedingly increased by a parabolic reflector behind it, or a lens in front of it. Such a light was brilliantly visible at 66 miles' distance.

(241.) Observations of the Angles. These should be repeated as often as possible. In extended surveys, three sets, of ten each, are recommended. They should be taken on different parts of the circle. In ordinary surveys, it is well to employ the method of "traversing," Art. (94). In long sights, the state of the atmosphere has a very remarkable effect on both the visibility of the signals and on the correctness of the observations.

When many angles are taken from one station, it is important to record them by some uniform system. The form given below is convenient. It will be noticed that only the minutes and seconds of the second vernier are employed, the degrees being all taken from the first:

Observations at ——————.

Station observed to.	READINGS.		Mean Reading.	Right or Left of Preceding Object.	Remarks.
	Vernier A.	Vernier B.			
A	70° 19′ 0″	18′ 40″	70° 18′ 50″		
B	103° 32′ 20″	32′ 40″	103° 32′ 30″	R.	
C	115° 14′ 20″	14′ 50″	115° 14′ 35″	R.	

When the angles are "repeated," the multiple arcs will be registered under each other, and the mean of the seconds shown by all the verniers at the first and last readings be adopted.

THE GREAT THEODOLITE, used on the Coast Survey for the observation of the angles in the primary triangles, has a horizontal circle thirty inches in diameter, graduated to five minutes, and reading to single seconds by three micrometer microscopes, placed 120° apart. The telescope has a focal length of four feet.

When the country over which the triangulation extends is flat, it has been found necessary to elevate the transit some distance from the surface of the ground, the stratum of air near the surface being so disturbed by exhalations and inequalities of temperature and density as to render accurate observation impossible. The plan adopted on the Coast Survey is as follows: On the top of a signal-tripod, forty-three feet high, is placed a cap-block, into which is mortised a square hole to receive the signal-pole. Around the tripod, but not touching it, is erected a rectangular scaffold, forty feet high. On the top of it is a platform, from which the observations are taken, the signal-pole being removed from the cap-block, and the transit placed so that its centre shall be precisely over the station-point.

(242.) **Reduction to the Centre.** It is often impossible to set the instrument precisely at or under the signal which has been observed. In such cases proceed thus: Let C be the centre of the signal, and R C L the desired angle, R being the right-hand object and L the left-hand one. Set the instrument at D, as near as possible to C, and measure the angle R D L. It may be less than R C L, or greater than it, or equal to it, according as D lies without the circle passing through C, L, and R, or within it,

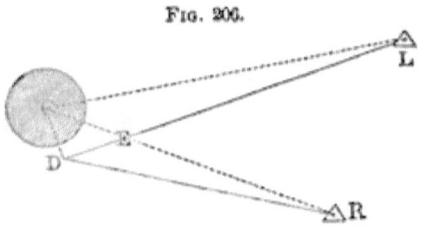

FIG. 206.

or in its circumference. The instrument should be set as nearly as possible in this last position. To find the proper correction for the observed angle, observe also the angle LDC (called the angle of direction), counting it from $0°$ to $360°$, going from the left-hand object toward the left, and measure the distance DC. Calculate the distances CR and CL with the angle RDL, instead of RCL, since they are sufficiently nearly equal. Then,

$$RCL = RDL + \frac{CD . \sin. (RDL + LDC)}{CR . \sin. 1''} - \frac{CD . \sin. LDC}{CL . \sin. 1''}$$

The last two terms will be the number of seconds to be added or subtracted. The trigonometrical signs of the sines must be attended to. The log. sin. $1'' = 4.6855749$. Instead of dividing by sin. $1''$, the correction without it, which will be a very small fraction, may be reduced to seconds by multiplying it by 206265.

Example.—Let $RDL = 32° 20' 18''.06$; $LDC = 101° 15' 32''.4$; $CD = 0.9$; $CR = 35845.12$; $CL = 29783.1$.

The first term of the correction will be $+ 3''.750$, and the second term $- 6''.113$. Therefore, the observed angle RDL must be diminished by $2''.363$, to reduce it to the desired angle RCL.

Much calculation may be saved by taking the station D so that all the signals to be observed can be seen from it. Then only a single distance and angle of direction need be measured.

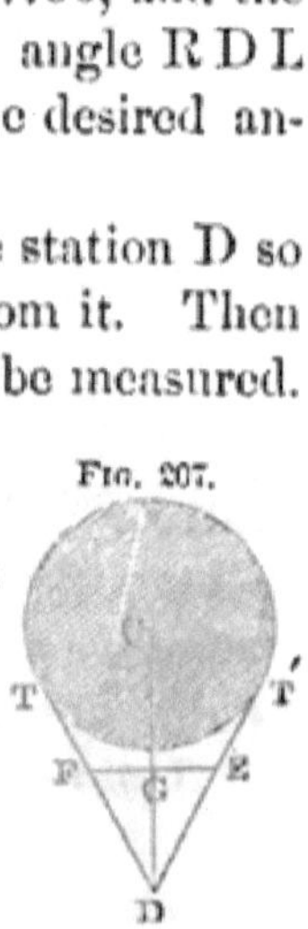

It may also happen that the centre, C, of the signal cannot be seen from D. Thus, if the signal be a solid circular tower, set the theodolite at D, and turn its telescope so that its line of sight becomes tangent to the tower at T, T'; measure on these tangents equal distances, DE, DF, and direct the telescope to the middle, G, of the line EF. It will then point to the centre, C; and the distance DC will equal the distance from D to the tower plus the radius obtained by measuring the circumference.

Fig. 207.

If the signal be rectangular, measure D E, D F. Take any point G on D E, and on D F set off D H $= DG \frac{DF}{DE}$. Then is G H parallel to E F (since D G : D H :: D E : D F), and the telescope directed to its middle, K, will point to the middle of the diagonal E F. We shall also have D C $= DK \frac{DE}{DG}$.

Fig. 208.

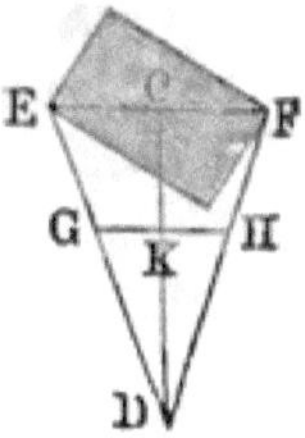

Any such case may be solved by similar methods.

The "*phase*" of objects is the effect produced by the sun shining on only one side of them, so that the telescope will be directed from a distant station to the middle of that bright side instead of to the true centre. It is a source of error to be guarded against. Its effect may, however, be calculated.

When the signal is a tin cone:

Let r = radius of the signal,
 Z = angle at the point of observation between the sun and the signal,
 D = the distance.

Then, the correction $= \pm \frac{r \cos.^2 \frac{1}{2} Z}{D \sin. 1''}$.

(243.) **The Angles.** The triangles observed are supposed to have sides of such length that the sum of the three angles exceeds 180° by a certain sensible quantity called the "*spherical excess.*" This is usually only a few seconds. For a triangle containing about 76 square miles, which, if equilateral, would have sides 13 miles long, the spherical excess is only one second. For a triangle with sides of 102 miles it is one minute.

It must be determined before we can know how much the error is, and therefore what the correct sum and correction should be.

(244.) The true spherical excess is found by this principle: "The surface of a spherical triangle is measured by the excess of its angles above two right angles multiplied by the trirectangular triangle."[1]

Hence the surfaces of spherical triangles are to each other as their respective spherical excesses.

Let s = surface of given triangle,
t = surface of trirectangular triangle,
e = spherical excess of given triangle,
e' = spherical excess of trirectangular triangle.

Then, we have:

$$s : t :: e : e'.$$

$t = \frac{1}{8}$ surface of sphere $= \frac{1}{8} \times 4\pi r^2 = \frac{1}{2}\pi r^2$.
$e' = (3 \times 90°) - 180° = 90°$.

Then, $s : \frac{1}{2}\pi r^2 :: e : 90°$.

Whence, $e = s \times \dfrac{648000''}{\pi r^2}$ in seconds.

s and r are in the same unit of measure.

The fraction is a constant quantity whose logarithm is $\overline{10}.6746069$, the mean radius of the earth being taken as 20888629 feet; the greater radius being 20923596, and the smaller radius 20853662.[2]

The surface s, being very small compared with r^2, may be obtained with sufficient accuracy for this object by treating the triangle as if it were plane.

Then, when two sides and the contained angle are given, we have:

$$s = \tfrac{1}{2} a b \cdot \sin. C.$$

When two angles and the included side are given, we have:

$$s = \tfrac{1}{2} a^2 \times \frac{\sin. B \cdot \sin. C}{\sin. (B + C)}.$$

Approximately, the spherical excess (in seconds) equals the area (in square miles) divided by 75.5.

---

[1] Davies's Legendre, Book IX., prop. 18. [2] According to Sir John Herschel.

Having found the spherical excess, if the sum of the angles of the triangle does not equal 180° plus this excess, the difference is distributed among them as in Art. (245).

(**245.**) **Correction of the Angles.** When all the angles of any triangle can be observed, their sum should equal 180° plus the "spherical excess." If not, they must be corrected. If all the observations are considered equally accurate, one-third of the difference of their sum from 180° plus the spherical excess, is to be added to or subtracted from each of them. But if the angles are the means of unequal numbers of observations, their errors may be considered to be inversely as those numbers, and they may be corrected by this proportion: As the sum of the reciprocals of each of the three numbers of observations is to the whole error, so is the reciprocal of the number of observations of one of the angles to its correction.

It is still more accurate, but laborious, to apportion the total error, or difference from 180° plus the spherical excess, among the angles inversely as the "*weights*."[1] On the United States Coast Survey, in six triangles measured in 1844 by Prof. Bache, the *greatest* error was six-tenths of a second.

(**246.**) **Interior Filling-up.** The stations whose positions have been determined by the triangulation are so many fixed points, from which more minute surveys may start and interpolate any other points. The trigonometrical points are like the observed latitudes and longitudes which the mariner obtains at every opportunity, so as to take a new departure from them, and determine his course in the intervals by the less precise methods of his compass and log. The chief interior points may be obtained by "secondary triangulation," and the minor details be then filled in by any of the methods of surveying, with chain, compass, or transit, already explained, or by the plane-table.

With the transit or theodolite, "traversing" is the best mode of surveying, the instrument being set at zero, and being

---

[1] L. S., Art. (369).

then directed from one of the trigonometrical points to another, which line therefore becomes the "Meridian" of that survey. On reaching this second point, in the course of the survey, and sighting back to the first, the reading should of course be 0°.

CHAPTER II.

CALCULATING THE SIDES OF THE TRIANGLES.

(247.) ONE side of a spherical triangle having been measured or calculated, and all the angles observed, the other sides can be computed by employing the principles of spherical trigonometry. This, however, is very laborious, and other methods have been adopted which, with less work, give results equally accurate.

(248.) **Delambre's Method.** Imagine the three angular points of each spherical triangle to be joined by straight lines, chords of the arc, so as to form a plane triangle, as in Fig. 209. Reduce the given curved side to its chord, and the spherical angles to the plane angles contained by these chords. Compute the other sides or chords by plane trigonometry, and then calculate the arcs corresponding to them.

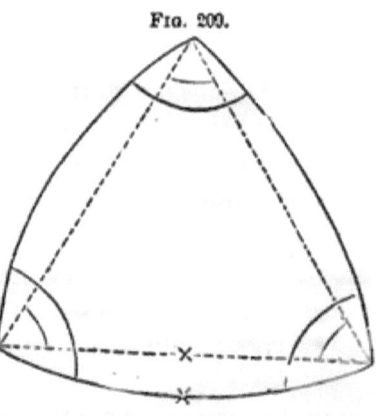

Fig. 209.

To reduce any arc to its chord we have:
Chord of an arc $a = 2 \sin. \frac{1}{2} a$.
Or, if a be the arc in terms of the radius:
Chord of $a = a - \frac{1}{24} a^3$.

To reduce an angle of a spherical triangle to the corresponding angle between the chords of the including arcs:

Let A B C, Fig. 210, be a spherical triangle, and O the centre of the sphere. It is required to reduce the spherical

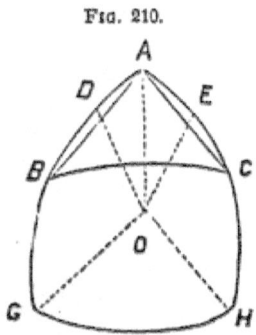

Fig. 210.

angle at A to the plane angle between the chords A B and A C. Draw O G and O H parallel to A B and A C, and prolong the arcs to G and H. The lines O D and O E, bisecting the chords A B and A C, will be perpendicular to them, and also to O G and O H. Then D G and E H are quadrants. Now, in the spherical triangle A G H, having the arcs A G and A H, and the included angle G A H, the measure of the angle G O H is found by spherical trigonometry. But G O H = B A C, the required angle.

The sum of the three plane angles thus found will be equal to two right angles, if the observations of the spherical angles and the work of reducing have been correctly done.

(249.) **Legendre's Method.** His theorem is this: "In any spherical triangle, the sides of which are very small compared to the radius of the sphere, if each of the angles be diminished by $\frac{1}{3}$ of the true spherical excess, the sines of these angles will be proportional to the lengths of the opposite sides; and the triangle may therefore be calculated as if it were a plane one."

This is the easiest method.

All three methods were used for the French "*Base du système métrique.*"

In the British "Ordnance Survey" the triangles were mostly calculated by the second method, and checked by the third.

The difference on 100 miles is only a fraction of a yard.

(250.) **Co-ordinates of the Points.** The *polar spherical co-ordinates* of a point with respect to another point are these:

the length of the arc of the great circle passing through the points, and its azimuth, i. e., the angle it makes with the meridian passing through one of its points.

The *rectangular spherical coördinates* of a point have for axes the meridian passing through the origin, and a perpendicular to it. For short distances these may be regarded as in one plane. For greater distances new meridians must be taken—say, not farther apart than 50 miles.

Within that limit the successive triangles may be conceived to be turned down into the same plane.

The astronomical coördinates of a point are its latitude and longitude. These are determined by practical astronomy.

The transformation of these coördinates to polar or rectangular, and *vice versa*, is very important. It is done by *spherical trigonometry*. The latitude and longitude of any one point are very accurately determined by the mean of a great number of astronomical observations, and those of the other points are calculated from these. Those of some other points may be observed as checks.

It is found that the observed and calculated latitude and longitude of a place do not always agree, even when the earth is considered as an ellipsoid of revolution; in consequence of the irregularity of the form of the earth. The difference of the "geodesic" from the astronomical determination of difference of latitude and longitude, is called the "station error."

A "geodesic line" is the shortest line which can be drawn on the ellipsoid, corresponding to an arc of a great circle on the sphere. It is the line of least curvature.

Fig. 211.

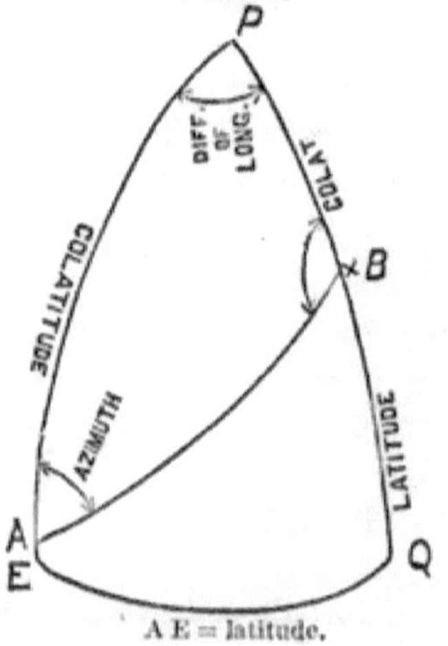

A E = latitude.

(251.) Prob. 1. Given latitude and longitude of A, and the azimuth and distance from A to B. Required the latitude and longitude of B, and the azimuth from B to A. The distance is measured on the arc of a great circle passing through those points, the earth being assumed to be a sphere.

We have given two sides and the included angle, to find the remaining parts.

By spherical trigonometry we have:

$$\tan. \tfrac{1}{2}(B+P) = \cot. \tfrac{1}{2}A \frac{\cos. \tfrac{1}{2}(AP-AB)}{\cos. \tfrac{1}{2}(AP+AB)}.$$

$$\tan. \tfrac{1}{2}(B-P) = \cot. \tfrac{1}{2}A \frac{\sin. \tfrac{1}{2}(AP-AB)}{\sin. \tfrac{1}{2}(AP+AB)}.$$

The azimuth from B to A $= B = \tfrac{1}{2}(B+P) + \tfrac{1}{2}(B-P)$.

The difference of long. $= P = \tfrac{1}{2}(B+P) - \tfrac{1}{2}(B-P)$.

To find the co-latitude of $B = PB$, we have:

$$\tan. \tfrac{1}{2}PB = \frac{\tan. \tfrac{1}{2}(AP-AB)\sin. \tfrac{1}{2}(B+P)}{\sin. \tfrac{1}{2}(B-P)}$$

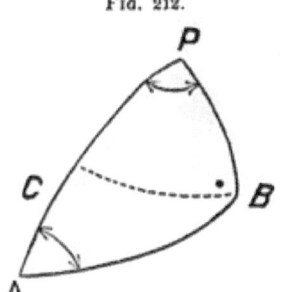

Fig. 212.

(252.) *Otherwise.*—Let fall from B, BC perpendicular to AP. Then,

$$\text{tang. } AC = \text{tang. } AB \cdot \cos. PAB.$$

$$PC = PA - AC.$$

$$\cos. PB = \frac{\cos. AB \cdot \cos. PC}{\cos. AC}.$$

$$\sin. APB = \frac{\sin. A \cdot \sin. AB}{\sin. PB}.$$

$$\sin. ABP = \frac{\sin. A \cdot \sin. AP}{\sin. PB}.$$

(253.) PROB. 2. Given latitude and longitude of A and B

CALCULATING THE SIDES OF THE TRIANGLES.

to find the distance between them and the azimuth from each to the other; i. e., to find the length and direction of the arc of a great circle passing through those points.

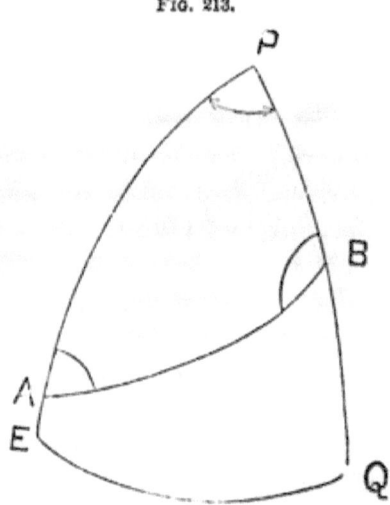

Fig. 213.

The angle of P is the difference of longitude, P B is the co-latitude of B, and P A is the co-latitude of A. Then we have two sides and the included angle to find the remaining parts.

$$\tan. \tfrac{1}{2}(B + A) = \cot. \tfrac{1}{2} P \frac{\cos. \tfrac{1}{2}(PA - PB)}{\cos. \tfrac{1}{2}(PA + PB)}.$$

$$\tan. \tfrac{1}{2}(B - A) = \cot. \tfrac{1}{2} P \frac{\sin. \tfrac{1}{2}(PA - PB)}{\sin. \tfrac{1}{2}(PA + PB)}.$$

$$B = \tfrac{1}{2}(B + A) + \tfrac{1}{2}(B - A).$$

$$A = \tfrac{1}{2}(B + A) - \tfrac{1}{2}(B - A).$$

$$\tan. \tfrac{1}{2} AB = \frac{\tan. \tfrac{1}{2}(PA - PB) \sin. \tfrac{1}{2}(B + A)}{\sin. \tfrac{1}{2}(B - A)}.$$

This is strictly a case of spherical location, required in planning a road between two distant points, and in navigating a vessel.

The distance may also be found thus:

Let a and β represent the co-latitudes of A and B.

$$\cos. AB = \cos. a \cdot \cos. \beta + \sin. a \cdot \sin. \beta \cdot \cos. P.$$

Put $\tang. \phi = \tan. a \cdot \cos. P$;

Then, $\cos. AB = \dfrac{\cos. a \cdot \cos.(\beta - \phi)}{\cos. \phi}.$

Any other sets of three parts of the triangle P A B being given, the rest can be found by spherical trigonometry.

(254.) For great accuracy, the earth must be regarded as a spheroid. The following formulas for computing the geodesic latitudes, longitudes, and azimuths of points of a triangulation, are from Captain Lee's Tables and Formulas:

Let K = distance in yards between two stations, the latitude and longitude of one of which are known, and u'' this same distance converted to second of arc.

L = latitude of first station.

M = longitude of first, + if west.

Z = azimuth of second station at first, counted from the south around, by the west, from 0° to 360°. The algebraic signs of the sine and cosine of this angle must be carefully attended to.

L', M', Z', the same things at second station, or quantities required.

a = the equatorial radius.

e = the eccentricity = 0.0817 = $\sqrt{\left(\frac{a^2 - b^2}{a^2}\right)}$.

R = the radius of curvature of the meridian, in yards.

N = the radius of curvature of a section perpendicular to the meridian, in yards.

$$u'' = \frac{K}{N \sin. 1''} = \frac{K(1 - e^2 \sin.^2 L)^{\frac{1}{2}}}{a \sin. 1''}.$$

$$L' = L - (1 + e^2 \cos.^2 L) u'' \cos. Z - (1 + e^2 \cos.^2 L) (u'' \sin. Z)^2 \tan. L \times \tfrac{1}{2} \sin. 1''.$$

$$M' = M + \frac{u'' \sin. Z}{\cos. L'}.$$

$$Z' = 180° + Z - \frac{u'' \sin. Z}{\cos. L'} \sin. \tfrac{1}{2} (L + L'), \text{ or}$$

$$Z' = 180° + Z - (u'' \sin. Z \tan. L + u''^2 \sin. Z \cos. Z \tfrac{1}{2} \sin. 1'').$$

CALCULATING THE SIDES OF THE TRIANGLE. 171

The quantity $\frac{u'' \sin. Z}{\cos. L'} \sin. \tfrac{1}{2}(L + L')$, or $(M' - M) \sin. \tfrac{1}{2} (L + L')$, by which the azimuth at one end of a line exceeds the azimuth at the other, is called the convergence of the meridians.

In terms of the coördinates of rectangular axes referred to one of the points of the triangulation, the latitude and longitude of which are known, y being the ordinate in the direction of the meridian, and x the ordinate perpendicular to it:

$$L' = L \pm \frac{y}{R \sin. 1''} - \tfrac{1}{2} \sin. 1'' \left(\frac{x}{N \sin. 1''}\right)^2 . \tan. \left(L \pm \frac{y}{R \sin. 1''}\right)$$

$$M' = M \pm \left(\frac{x}{N \sin. 1''}\right) \times \frac{1}{\cos. L'}.$$

$$Z' = 270° \pm \frac{x}{N \sin. 1''} \tan. L'.$$

THE END.

www.ingramcontent.com/pod-product-compliance
Lightning Source LLC
Chambersburg PA
CBHW020243170426
43202CB00008B/206